Health Benefits Of
Understanding Blood Types, Foods *and* Allergies

·········• ·········

Dr. Gideon Nnabuike,
DNP, PMHNP-BC, APRN, PGMS, MS, BSN.

Health Benefits of Understanding Blood Types, Foods, and Allergies

Copyright ©2024 Dr. Gideon Nnabuike

Paperback ISBN: 978-1-957809-97-7

All rights reserved. No part of this publication may be reproduced, distributed, or transmitted in any form or by any means, including photocopying, recording, or other electronic or mechanical methods without the prior written permission of the author except in the case of brief quotations embodied in reviews and certain other non-commercial uses permitted by copyright law.

Published by Cornerstone Publishing
A Division of Cornerstone Creativity Group LLC
Info@thecornerstonepublishers.com
www.thecornerstonepublishers.com

Author's Contact
To book the author to speak at your next event or to order bulk copies of this book, please, use the information below:
gidnna@gmail.com

Printed in the United States of America.

CONTENTS

PART ONE
UNDERSTANDING BLOOD TYPES AND THEIR SIGNIFICANCE

1. Blood Types: What Letters, Positive, And Negative Signs Mean 3
2. The Significance Of Blood Type 13
3. Your Blood Type And Foods 25

PART TWO
ALLERGIES

4. Overview Of Food Allergies 49
5. Diagnosing Food Allergies 53
6. Managing Food Allergies 59
7. Special Populations 69

PART THREE
FOODS, BENEFITS & DAILY RECOMMENDATIONS

8. Types Of Foods, Possible Side Effects, And Quantities Recommended 83

APPENDIX 175
REFERENCES 194

PART ONE
Understanding Blood Types and Their Significance

Recent research suggests that blood type may influence susceptibility to certain diseases, pregnancy complications, and even severe COVID-19 outcomes. However, these links require further investigation.

While we may all bleed the same, the type of blood that runs through our vessels is a unique aspect of our biology. Understanding blood types is critical for many reasons, including the foods you eat and blood transfusions, offering insights into human genetics and health.

Chapter One
BLOOD TYPES: WHAT LETTERS, POSITIVE, AND NEGATIVE SIGNS MEAN

Your blood type is determined by specific antigens present or absent on the surface of your red blood cells. Antigens can trigger the immune system to produce antibodies, which are proteins that attack perceived foreign invaders. Knowing your blood type is essential, especially if you need a blood transfusion, as incompatible transfusions can have serious health consequences.

TYPES OF BLOOD AND THEIR SIGNIFICANCE

The ABO system classifies blood into four major types: A, B, AB, and O. These are further categorized by the presence (positive) or absence (negative) of the Rh(D) antigen, resulting in eight major blood types:

- A+
- A-

- B+
- B-
- AB+
- AB-
- O+
- O-

A and B antigens are sugars, and the type of sugar antigens determines whether someone has A, B, AB, or O blood. The Rh factor is a protein antigen, with about 85% of the population being Rh-positive.

The International Society of Blood Transfusion has identified 45 different blood group systems, which include hundreds of foreign antigens.

RARE AND COMMON BLOOD TYPES IN THE UNITED STATES

Rare Blood Types (found in less than 5% of the population):

- AB-: 0.6%
- B-: 1.5%
- AB+: 3.4%

Common Blood Types (found in more than 70% of the population):

- O+: 37.4%
- A+: 35.7%

WHAT IS GOLDEN BLOOD?

Golden blood, the rarest known type, has no Rh antigens at all (Rhnull). It can be donated to almost anyone with Rh blood types, but those with golden blood can only receive the same type. Only about 50 people are known to have this blood type, first detected in Australian aboriginal people.

REASONS TO KNOW YOUR BLOOD TYPE

1. **Diet and Health:** Certain diets may be recommended based on blood type.

2. **Blood Transfusions:** Knowing your blood type ensures you receive compatible blood during surgery, injury, or illness, preventing hemolytic transfusion reactions.

3. **Donations:** Being aware of your blood type can help others, especially when supplies of your blood type are low.

Blood types also influence the likelihood of developing certain conditions, such as kidney stones, high blood pressure during pregnancy, and bleeding disorders. For instance, people with blood group A may be more susceptible to COVID-19 than those with blood group O.

COMPATIBILITY OF DIFFERENT BLOOD TYPES

Blood compatibility depends on whether the recipient's antibodies can attack the donor blood antigens. Here is a compatibility chart:

Your Blood Type	Compatible Donor Blood Types
O+	O+, O-

Your Blood Type	Compatible Donor Blood Types
O-	O-
A+	A+, A-, O+, O-
A-	A-, O-
B+	B+, B-, O+, O-
B-	B-, O-
AB+	AB+, AB-, A+, A-, B+, B-, O+, O-
AB-	AB-, A-, B-, O-

UNIVERSAL DONORS AND RECIPIENTS

Type O negative is the universal donor, while type AB positive is the universal recipient. However, blood compatibility testing (crossmatching) is performed before transfusions to ensure safety.

TESTING BLOOD TYPES IN PREGNANCY

Identifying Rh blood type in pregnancy is crucial to prevent Rh incompatibility, which can lead to severe consequences for Rh-positive fetuses born to Rh-negative mothers. If a pregnant woman is Rh-negative and the fetus is Rh-positive, she may develop antibodies that can attack future Rh-positive pregnancies. Rh-negative pregnant women are given RhoGAM or intravenous 'WinRho' to prevent antibody development.

HOW TO FIND OUT YOUR BLOOD TYPE

A blood test can determine your blood type. Blood typing is free if you donate blood or plasma. It's not part of routine blood tests but is ordered for surgeries, transfusions, organ transplants, or pregnancy. You can request a blood type test from your healthcare provider, though insurance may not cover it if it's not medically necessary. Home blood type tests are available and generally accurate if performed correctly.

Blood typing using the ABO system and Rh factor results in eight major blood types. Compatibility is crucial for transfusions and pregnancy to avoid severe reactions. Professional laboratory blood typing is more reliable than home tests, though home tests are available.

Understanding blood types is essential for personal health and the safety of medical procedures.

HOW YOUR BLOOD TYPE CAN AFFECT YOUR HEALTH

How Blood Type is Determined

Blood type, like eye color, is inherited from parents and is determined by the presence or absence of specific antigens. If these antigens are unfamiliar to the body, they can elicit an immune reaction.

ABO and the Most Common Blood Types

The ABO blood group system classifies blood types based on the presence of different antigens on red blood cells and antibodies in the plasma. Along with the RhD antigen status, these factors determine compatibility for safe blood transfusions.

Group A

The surface of red blood cells contains A antigen, and the plasma has anti-B antibodies, which attack blood cells with B antigen.

Group B

The surface of red blood cells contains B antigen, and the plasma contains anti-A antibodies, which attack blood cells with A antigen.

Group AB

Red blood cells have both A and B antigens, while the plasma does not contain anti-A or anti-B antibodies. Thus, people with type AB blood can receive any ABO blood type.

Group O

The plasma contains both anti-A and anti-B antibodies, but the red blood cells lack A and B antigens. This blood type can be given to any ABO blood type.

Rhesus Factor

The Rh factor is a specific protein found on red blood cells. If present, the blood type is Rh-positive; if absent, it is Rh-negative.

Blood Testing

A blood test can determine an individual's blood type by mixing blood with substances that contain A or B antibodies or Rh factor and observing the reactions.

HOW BLOOD TYPE AFFECTS HEALTH

Heart Disease

Research indicates that people with type O blood have a lower risk of coronary heart disease. This might be due to lower cholesterol levels and reduced amounts of clotting-related proteins in other blood types.

Memory

A small study found that individuals with AB blood type are more likely to develop cognitive and memory problems, potentially leading to dementia.

Life Span

People with type O blood tend to live longer, likely due to a lower risk of heart and blood vessel diseases.

Stomach Cancer

Individuals with blood types A, AB, and B are at higher risk for stomach cancer, particularly those with type A blood, possibly due to a higher prevalence of H. pylori infection.

Stress

People with type A blood have higher cortisol levels, the stress hormone, making it harder for them to manage stress.

Pancreatic Cancer

The risk of pancreatic cancer is higher for those with types A, AB, or B blood due to molecules in type A and B red blood cells that help H. pylori bacteria grow.

Malaria

Individuals with type O blood have a reduced risk of malaria because the parasite cannot attach to their cells.

Ulcers

Peptic ulcers occur more frequently in people with type O blood.

Diabetes

Type 2 diabetes is more common in people with blood types A and B, though further research is needed to understand why.

Fertility

Women with type O blood may have fewer healthy eggs, but more research is required to understand this fully.

Rheumatic Diseases

A study showed that certain rheumatic diseases are more common in people with blood types A and O.

Blood Clots

People with blood types A, B, or AB are at a higher risk of venous thromboembolism (VTE), a condition where blood clots form in deep veins.

Stroke

Individuals with type AB blood have a higher risk of stroke due to a greater tendency to clot.

Lupus

Lupus, an autoimmune disease, tends to cause more severe symptoms in people with blood types A and B.

Multiple Sclerosis (MS)

Those with blood types A and B have a higher risk of developing MS, a condition where the immune system attacks the protective layer around nerves.

Inflam matory Bowel Disease (IBD)

Studies have shown that people with type O blood are less likely to develop Crohn's disease, a type of IBD.

Hemorrhage

Individuals with type O blood are more prone to fatal hemorrhages following severe trauma, though the reasons for this require further investigation.

RISKS IN PREGNANCY

During pregnancy, the Rh antigen becomes relevant when a mother with Rh-negative blood carries a fetus with Rh-positive blood. Antibodies from the Rh-negative mother can potentially harm the red blood cells of her Rh-positive unborn infant by entering their bloodstream.

UNDERSTANDING THE CONNECTION BETWEEN BLOOD TYPE AND DISEASES

Researchers are still exploring the relationship between blood type and certain diseases. It is believed that a combination of genetic predisposition and environmental factors contributes to the onset of numerous conditions. Additional research is necessary to gain a comprehensive understanding of this association.

Chapter Two

THE SIGNIFICANCE OF BLOOD TYPE

You wouldn't know it from looking at a person, but inside the blood flowing through our veins are tiny variations that categorize every human into one of these blood-type groups: A-positive, A-negative, B-positive, B-negative, O-negative, O-positive, AB-positive, and AB-negative. Often, these minute differences don't matter until they do—such as when you're in the hospital needing a blood transfusion or after you've donated blood and learned your type. Some people find out during pregnancy when special treatment is required for someone with a negative blood type.

IMPORTANCE OF KNOWING YOUR BLOOD TYPE

Knowing your blood type can be crucial in an emergency and provide critical insight into your health. Ongoing research into blood types suggests they may matter more than we give them credit for, particularly when assessing risks for certain health conditions, especially heart disease. These invisible blood differences may affect one's susceptibility to cardiovascular problems.

WHAT DOES BLOOD TYPE MEAN?

The letters A, B, and O represent various forms of the ABO gene, which program our blood cells to form different blood groups. If you have type AB blood, your body produces A and B antigens in red blood cells. A person with type O blood doesn't make any antigens. Blood is classified as "positive" or "negative" based on the presence of proteins on red blood cells. If your blood has these proteins, you're Rhesus (Rh) positive.

UNIVERSAL DONORS

People with type O-negative blood are considered "universal donors" because their blood lacks antigens and proteins, making it compatible with anyone in an emergency.

THE EVOLUTION OF BLOOD TYPES

Researchers don't fully understand why different blood types exist. Factors such as ancestral origins and past infections that spurred protective mutations in the blood may have contributed to this diversity. For example, people with type O blood may get sicker with cholera, while those with type A or B blood may be more prone to blood clotting issues.

BLOOD TYPES AND HEART DISEASE

People with type A, B, or AB blood are more likely than those with type O to have a heart attack or experience heart failure. According to the American Heart Association, people with type A and B blood have a higher risk of developing deep vein thrombosis and pulmonary embolism, severe blood clotting disorders that can increase the risk of heart failure. This increased risk may be due to inflammation caused by proteins in type A and B blood, leading to more blockages in veins and arteries.

BLOOD TYPES AND COVID-19

There is anecdotal evidence suggesting that people with type O blood may have a lower risk of severe COVID-19 disease because the virus binds to cells differently based on blood type.

OTHER HEALTH IMPLICATIONS OF BLOOD TYPES

1. **Type O Blood**: Lower risk of heart disease and blood clotting but higher susceptibility to hemorrhaging and bleeding disorders, especially after childbirth.

2. **Type AB Blood**: Increased risk of cognitive impairment, including trouble remembering, focusing, or making decisions.

3. **Type A Blood**: More cortisol, making it harder to deal with stress, and higher risk for certain cancers like stomach and pancreatic cancer.

4. **Type B Blood**: Higher risk of pancreatic cancer due to molecules that help H. pylori bacteria grow.

SHOULD YOU CHANGE YOUR LIFESTYLE BASED ON BLOOD TYPE?

While blood type can influence the risk of certain health conditions, major factors like diet, exercise, and environmental exposure play more significant roles in determining overall health. Maintaining a heart-healthy diet that lowers inflammation is recommended for everyone, regardless of blood type. Future research may provide more tailored medical advice based on blood type.

BLOOD TYPING AND ITS IMPORTANCE

Blood typing includes identifying A and B sugar antigens and Rh proteins on the surface of red blood cells. The presence or absence of Rh proteins is significant during pregnancy or when a blood transfusion is needed. Rh-negative pregnant women should take precautions against developing anti-Rh antibodies to protect the health of future pregnancies.

BLOOD TYPING IN PREGNANCY

Blood testing is part of early prenatal care and includes blood typing, RBC count, WBC count, platelet count, and electrolyte levels. The Rh type is inherited separately from the ABO type, with most people being Rh-positive. Rh incompatibility can complicate pregnancy for an Rh-negative pregnant person carrying an Rh-positive fetus. Rh sensitization occurs when blood from an Rh-positive fetus enters the bloodstream of the Rh-negative mother, stimulating the production of anti-Rh antibodies, which can harm future pregnancies.

RH FACTOR OF EACH PARENT

Having Rh factor protein on the surface of RBCs is a dominant trait. If one of a baby's genetic parents is Rh positive, the child may be Rh positive, even if the other parent is Rh negative. The Rh factor can affect pregnancy in various ways depending on the Rh status of both parents.

Other Parent Rh Negative	Other Parent Rh Positive
Pregnant Person Rh Negative	No incompatibility
Pregnant Person Rh Positive	No incompatibility

If the pregnant person is carrying a fetus conceived through in vitro fertilization (such as in surrogacy or egg donation), the Rh factor can also affect the pregnancy.

RH FACTOR INCOMPATIBILITY, ANTIBODIES, AND FETAL RISKS

During pregnancy, the Rh antigen becomes relevant if the mother is Rh-negative and the fetus is Rh-positive. Antibodies from an Rh-negative mother can potentially harm the red blood cells of her Rh-positive unborn infant by entering their bloodstream.

BLOOD TYPING DURING PREGNANCY

Blood typing is essential during pregnancy because Rh incompatibility can lead to severe problems for the fetus. Rh incompatibility occurs when the pregnant person is Rh-negative, and the fetus is Rh-positive, potentially causing an immune-driven hemolytic reaction in the fetus's blood. Hemolysis, the breaking down of red blood cells (RBCs), can result in anemia and organ failure in the fetus or baby. The severity of the hemolytic reaction varies, with some fetuses surviving mild anemia if treated early, while severe cases can lead to lifelong disability or death.

WHEN DO RH ANTIBODIES FORM?

Rh-negative pregnant individuals can produce antibodies against the Rh proteins in the fetus's red blood cells. Usually, the fetus's blood does not mix with the pregnant person's blood until labor and delivery. However, antibody formation can occur after invasive prenatal testing procedures like amniocentesis, as well as after ectopic pregnancy, miscarriage, or pregnancy termination if there is Rh incompatibility. Since antibody formation takes time, it usually does not affect the first pregnancy where sensitization occurs. The Rh factor does not develop on the fetus's red blood cells until about eight weeks of pregnancy, making early pregnancy loss or termination less risky.

PREVENTING ANTI-RH ANTIBODY FORMATION

If you are Rh-negative, you may need medication to prevent anti-Rh antibody formation during and after pregnancy. The medication, RhoGAM or Rho(D) immune globulin, is necessary to prevent the development of antibodies if you are carrying a fetus conceived with either your egg cell or a donor egg cell.

USE OF RHOGAM

Rh-negative pregnant individuals who have not formed any Rh antibodies will receive RhoGAM around 28 weeks of pregnancy to prevent antibody formation. Injections may be scheduled sooner if invasive procedures are planned or if there has been a pregnancy loss or termination. If the baby is found to be Rh-positive at birth or is not tested for Rh factor, an additional dose of RhoGAM will be given within 72 hours after delivery.

WHEN RH ANTIBODIES ARE PRESENT

If Rh antibodies are already present due to a previous Rh-incompatible pregnancy, the current pregnancy will be considered high risk. Early delivery might be recommended to reduce the likelihood of hemolysis, and the baby's blood levels will be monitored for the need for a blood transfusion.

BLOOD TRANSFUSIONS AND RH-NEGATIVE BLOOD

When donor blood is used for a transfusion, it is typed and matched for compatibility, including matching the Rh type and ABO type. If an Rh-negative person receives Rh-positive blood, and they do not have pre-existing Rh antibodies, the transfusion will not affect them immediately but may lead to the development of Rh antibodies. Once Rh antibodies are present, future transfusions of Rh-positive blood may result in a hemolytic transfusion reaction and affect any Rh-positive fetus in subsequent pregnancies.

COMMUNICATING RH FACTOR STATUS DURING PREGNANCY

Pregnant individuals can expect to have Rh typing during the first trimester. If you are Rh-positive, no action is needed. If you are Rh-negative, treatment to prevent the formation of Rh antibodies may be required.

FUTURE HEALTH OF BABIES IN RH-INCOMPATIBLE PREGNANCIES

Rh-negative fetuses and babies are not at risk regardless of the Rh type of the pregnant person. However, Rh-positive fetuses carried by Rh-negative individuals are at higher risk after the first pregnancy. With proper screening and the use of RhoGAM, fewer Rh-negative pregnant individuals develop anti-Rh antibodies that can affect future pregnancies. If an antibody is present, the fetus's health will be closely monitored for signs of hemolytic disease.

In cases of mild anemia, the fetus may not need treatment before or after birth. For moderate anemia, early delivery may be recommended, followed by potential blood transfusions or treatment for jaundice. Severe anemia may require in-utero blood transfusions and possibly a cesarean section. Post-birth, the baby may need further transfusions and treatment for jaundice to prevent severe complications like hydrops and kernicterus.

Blood type, defined by the presence or absence of the Rh factor, is critical during pregnancy and before blood transfusions. Rh-negative individuals should avoid receiving Rh-positive blood. Rh incompatibility occurs when an Rh-negative pregnant individual carries an Rh-positive fetus, leading to the formation of anti-Rh antibodies, especially during labor and delivery. Preventive treatment with RhoGAM can mitigate the risks associated with Rh incompatibility, protecting future pregnancies from severe health consequences.

BLOOD TYPES AND WHAT YOU CAN EAT

Your blood type may influence the best diet for you. Here are dietary recommendations for each blood type:

Type A Blood

- **Diet:** A meat-free diet based on fruits, vegetables, beans, legumes, and whole grains.
- **Recommendation:** Ideally, consume organic and fresh foods because people with type A blood have a sensitive immune system.

Type B Blood

- **Avoid:** Corn, wheat, buckwheat, lentils, tomatoes, peanuts, sesame seeds, and chicken.
- **Recommendation:** Eat green vegetables, eggs, certain meats, and low-fat dairy.

Type AB Blood

- **Focus on:** Tofu, seafood, dairy, and green vegetables.
- **Avoid:** Caffeine, alcohol, and smoked or cured meats.
- **Note:** People with type AB blood tend to have low stomach acid.

Type O Blood

- **Diet:** A high-protein diet heavy on lean meat, poultry, fish, and vegetables.
- **Limit:** Grains, beans, and dairy.
- **Supplements:** Various supplements may help with digestive issues common in people with type O blood.

RAREST AND MOST COMMON BLOOD TYPES

What is the Rarest Blood Type?

- **Rarest Blood Type:** AB negative.
- **Statistics:**
    - Occurs in 1% of Caucasians
    - 0.3% of African Americans
    - 0.1% of Asians
    - 0.2% of Latino Americans

What is the Most Common Blood Type?

- **Most Common Blood Type:** O positive.
- **Statistics:**
    - Occurs in 37% of Caucasians
    - 47% of African Americans
    - 39% of Asians
    - 53% of Latino Americans

UNIVERSAL DONOR AND RECIPIENT BLOOD TYPES

Universal Donor

- **Blood Type:** O negative.
- **Significance:** Can be used for transfusions in emergencies for any blood type.
- **Statistics:** Only 7% of the U.S. population has O-negative blood:
    - 8% of Caucasians
    - 4% of African Americans

- 1% of Asians
- 4% of Latino Americans

Universal Recipient
- **Blood Type:** AB positive.
- **Significance:** Can receive blood from any blood type.

SUMMARY

Blood type is determined by the presence or absence of specific antigens that can trigger an immune response. While everyone's blood contains these tiny variations, they can significantly impact diet recommendations, health risks, and the compatibility of blood transfusions. Understanding your blood type can offer critical insights into your health and guide dietary and medical decisions.

Chapter Three

YOUR BLOOD TYPE AND FOODS

BLOOD TYPE, ANTIGENS, ANTIBODIES, AND GENOTYPE

Blood Type: What Do Blood Types Mean?

- **Blood Group A:** Has 'A' antigens on the red blood cells and anti-B antibodies in the plasma.
- **Blood Group B:** Has 'B' antigens on the red blood cells and anti-A antibodies in the plasma.
- **Blood Group O:** Has no antigens on the red blood cells but has both anti-A and anti-B antibodies in the plasma.
- **Blood Group AB:** Has both 'A' and 'B' antigens on the red blood cells but no antibodies in the plasma.

WHAT IS AN ANTIGEN IN SIMPLE TERMS?

An antigen is any substance that causes your immune system to produce antibodies against it. This means your immune system does not recognize the substance and is trying to fight it off. Examples of antigens include chemicals, bacteria, viruses, or pollen.

WHAT IS AN ANTIBODY'S SIMPLE DEFINITION?

Antibodies are proteins produced by your immune system to protect you when an unwanted substance enters your body. They bind to these substances to eliminate them from your system. Another term for antibody is immunoglobulin.

WHAT IS A GENOTYPE'S SIMPLE DEFINITION?

A genotype is the genetic makeup of an individual organism, which determines or contributes to its phenotype. The contrasting terms genotype and phenotype define an organism's characteristics or traits.

THE MOST COMMON AND RAREST BLOOD TYPES

What is the Most Common Blood Type? The most common blood type is O positive, which is the blood type for 37% of the population.

What is the Rarest Blood Type? The rarest blood type is AB negative, occurring in:

- 1% of Caucasians
- 0.3% of African Americans
- 0.1% of Asians
- 0.2% of Latino Americans

WHAT IS 'GOLDEN BLOOD'?

Golden Blood: Rh-null is the rarest blood type and is considered "golden blood" due to a complete lack of antigens in the Rh system. Only 43 people worldwide have been recorded with this blood type, with only nine active donors.

RHESUS BLOOD GROUP SYSTEM

Description:
- A blood test classifies blood according to the presence or absence of the Rhesus (Rh) antigen.
- Rh antigen is lacking in the red blood cells of 15% or less of the population, known as Rh-negative.
- An Rh-negative mother and an Rh-positive father are at risk of having a baby with hemolytic disease of the newborn (HDN).
- An Rh immunoglobulin (RhIg) injection is given to an Rh-negative mother at 28 weeks gestation and repeated after delivery if the baby is Rh-positive.

Purpose:
- To classify blood according to the presence or absence of the Rh antigen.
- To screen prospective blood donors for the weak D antigen.
- To prevent HDN in pregnancy or a reaction after a transfusion.

Preparation and Procedure:
- No dietary restrictions are required.
- Blood tests are performed at the first prenatal visit to determine Rh sensitization status.
- The test involves a blood sample, causing slight discomfort from the needle puncture.
- Standard precautions and proper labeling procedures are maintained to avoid complications such as hematoma.

Interpretation:
- Normal results: Presence of D antigen indicates Rh-positive; absence indicates Rh-negative.
- Abnormal results: Rh incompatibility, a common and severe cause of HDN, can occur when an Rh-negative woman and an Rh-positive man have an Rh-positive baby.

BLOOD TYPES AND WHAT YOU CAN EAT

Type A Blood:
- **Diet:** Meat-free diet based on fruits, vegetables, beans, legumes, and whole grains. Prefer organic and fresh foods due to a sensitive immune system.

Type B Blood:
- **Avoid:** Corn, wheat, buckwheat, lentils, tomatoes, peanuts, sesame seeds, and chicken.
- **Recommended:** Green vegetables, eggs, certain meats, and low-fat dairy.

Type AB Blood:
- **Focus on:** Tofu, seafood, dairy, and green vegetables.
- **Avoid:** Caffeine, alcohol, and smoked or cured meats. People with type AB blood tend to have low stomach acid.

Type O Blood:
- **Diet:** High-protein diet heavy on lean meat, poultry, fish, and vegetables.
- **Limit:** Grains, beans, and dairy.
- **Supplements:** Various supplements may help with digestive issues common in people with type O blood.

BLOOD TYPE COMPATIBILITY FOR TRANSFUSIONS

Compatibility of Different Blood Types:
- **O+:** Compatible with O+ and O-.
- **O-:** Compatible with O- only.
- **A+:** Compatible with A+, A-, O+, and O-.
- **A-:** Compatible with A- and O-.
- **B+:** Compatible with B+, B-, O+, and O-.
- **B-:** Compatible with B- and O-.
- **AB+:** Compatible with all blood types.
- **AB-:** Compatible with AB-, A-, B-, and O-.

THE IMPORTANCE OF KNOWING YOUR BLOOD TYPE

Knowing your blood type is essential for:

- Receiving the right type of blood during a transfusion.
- Preventing severe reactions in transfusions.
- Managing Rh factor during pregnancy to prevent hemolytic disease in the newborn.
- Understanding potential health risks associated with your blood type.

Methods to Identify Your Blood Type:

1. **Doctor's Blood Test:** Involves drawing blood and performing forward and reverse typing to determine blood type.

2. **Donating Blood:** Blood type is identified when you donate blood, although results are not immediate.

3. **At-Home Blood Test:** Involves pricking your finger, applying blood to a test card, and comparing the results.

SHOULD BLOOD TYPE INFLUENCE YOUR DIET?

Research suggests that different blood types may benefit from specific diets. While it's essential to follow general healthy eating guidelines, considering blood type-specific recommendations may offer additional health benefits.

BLOOD GROUP O POSITIVE: CHARACTERISTICS, BENEFITS, AND DIETARY RECOMMENDATIONS

Biological Composition: Blood group O positive (O+) has Rhesus factor D positive. It has no antigens on the red blood cells but has both anti-A and anti-B antibodies in the plasma. Individuals with blood group O have defective A or B transferases, leaving H antigen, the terminal sugar of which is fucose, on the RBC. The base structure of the ABO blood group system is the fucosyl galactose H-antigen, which is attached to glycoproteins and lipids on the surface of RBCs and other tissues.

Ethnicity & Population: Approximately 45% of Caucasians, 51% of African Americans, and 57% of Hispanics are type O. The demand for blood types O negative and O positive is high due to their compatibility for transfusions.

Benefits:

- O-positive individuals can donate blood to any positive blood type.
- They have a lower risk of heart diseases and memory problems, including dementia.
- Type O has the lowest risk for heart attacks and blood clots in the legs and lungs.

Foods to Avoid:

- Avoid grains, beans, and legumes.
- Avoid wheat, corn, and dairy for weight loss.
- Limit or avoid capers, cinnamon, cornstarch, corn syrup, nutmeg, black and white pepper, vanilla, and vinegar.

- Avoid caffeine and alcohol, as they can increase adrenaline levels.
- Fizzy drinks, tea, coffee, wine, beer, and spirits should be restricted.

Recommended Foods:
- High-protein foods such as meat, vegetables, fish, and fruit.
- Seafood, kelp, red meat, broccoli, spinach, and olive oil for weight loss.
- Fruits like plums, prunes, figs, grapefruit, and most berries.
- Lean meat and seafood for protein.

Diseases More Common in Type O Positive:
- Familial Mediterranean fever, stomach ulcers, systemic lupus erythematosus, systemic sclerosis, and Sjögren's syndrome.
- Lower risk of heart attack due to coronary artery disease compared to other blood types.
- Potential lower risk of COVID-19 infection, although more research is needed.

BLOOD GROUP O NEGATIVE: CHARACTERISTICS, BENEFITS, AND DIETARY RECOMMENDATIONS

Biological Composition: Blood group O negative (O-) has no Rhesus factor D. It lacks all highly reactive antigens. The absence of A and B antigens and the Rh factor means nothing alerts the recipient's immune system.

Ethnicity & Population: O-negative blood type is found in 8%

of Caucasians, 4% of African Americans, 1% of Asians, and 4% of Latino Americans. It is not extremely common, with only 7% of the U.S. population having O-negative blood.

Benefits:
- O-negative is the universal blood type, used in emergencies when the blood type is unknown.
- O-negative individuals can only receive O-negative blood.
- It is the most common blood type used for transfusions in trauma, emergency, and surgery.

Foods to Avoid:
- Dairy products like milk, cheese, and yogurt.
- Wheat products like bread, pasta, and cereals.
- Legumes like beans, lentils, and peanuts.
- Processed foods like fast food, packaged snacks, and sugary drinks.
- Avoid wheat, barley, corn, rice, potato, cauliflower, lentil, cabbage, kidney beans, alcohol, caffeine, spinach, mushrooms, oranges, kiwi, strawberry, blackberry, coconut, green peas, and peanut butter.

Recommended Foods:
- High-protein foods such as meat, vegetables, fish, and fruit.
- For weight loss: seafood, kelp, red meat, broccoli, spinach, and olive oil.

Diseases More Common in Type O Negative:
- Inflammation from dairy products and wheat, leading to weight gain and immune system imbalances.

- Increased risk of thyroid disease due to overactive immunity.
- Lower risk of heart disease and blood clotting but higher susceptibility to hemorrhaging or bleeding disorders.
- Higher risk of postpartum blood loss in women.
- Rh incompatibility during pregnancy, requiring RhoGAM injection.

BLOOD GROUP A POSITIVE: CHARACTERISTICS, BENEFITS, AND DIETARY RECOMMENDATIONS

Biological Composition: Blood group A positive (A+) has A antigens and the Rh factor on the red blood cells with anti-B antibodies in the plasma. The A allele encodes a glycosyltransferase that produces the A antigen (N-acetyl-galactosamine is its immunodominant sugar).

Ethnicity & Population: Approximately 32% of the U.S. population has A+ blood. It is one of the most transfused blood types and is highly beneficial for platelet and whole blood donations.

Benefits:
- Type A+ is in high demand for treating cancer patients and premature babies.
- A+ red blood cells can be given to both A+ and AB+ patients.
- Plasma and platelet donations from A+ individuals are crucial for various patients.

Foods to Avoid:
- Beef, pork, lamb, cow's milk, potatoes, yams, sweet potatoes.
- Certain vegetables like cabbage, eggplant, tomatoes, peppers, and mushrooms.
- Lima beans, melons, oranges, strawberries, mangos.
- Poultry other than chicken and turkey, fish like bluefish, barracuda, haddock, herring, catfish.
- Certain grains like wheat bran, multigrain bread, durum wheat, refined sugar, refined carbohydrates.
- Oils other than olive oil, artificial ingredients, and most condiments.

Recommended Foods:
- Soy protein, such as tofu.
- Certain grains like spelled, hulled barley, and sprouted bread.
- Walnuts, pumpkin seeds, and peanuts.
- Olive oil, fruits like blueberries and elderberries.
- Beans and legumes, dark leafy greens like kale, Swiss chard, and spinach.
- Garlic and onions, cold-water fish like sardines and salmon.
- Limited amounts of chicken and turkey, green tea, ginger.

Diseases More Common in Type A Positive:
- Higher risk of cancer, diabetes, anxiety disorders, and cardiovascular diseases.
- Increased risk of stomach cancer due to higher prevalence of

H. pylori infection, a bacterium commonly found in the stomach, which can cause inflammation and ulcers.

BLOOD GROUP A NEGATIVE: CHARACTERISTICS, BENEFITS, AND DIETARY RECOMMENDATIONS

Biological Composition: Blood group A negative (A-) has A antigens on the red blood cells with anti-B antibodies in the plasma but lacks the Rh factor. The A allele encodes a glycosyltransferase that produces the A antigen.

Ethnicity & Population: About 8% of the U.S. population has A- blood. This blood type is rare and partially due to the inheritance pattern of Rh- being a recessive trait.

Benefits:

- A- red blood cells can treat around 40% of the population.
- A- platelets are universal, meaning they can be given to patients of any blood group, making them crucial in medical treatments.

Foods to Avoid:

- Bananas, coconut, papaya, cashews, pistachios.
- Beer, chicken, fish, and eggs.

Recommended Foods:

- Similar to A+, including soy protein, certain grains, nuts and seeds, olive oil, certain fruits and vegetables, garlic, onions, and cold-water fish.
- Limited amounts of chicken and turkey, green tea, and ginger.

Diseases More Common in Type A Negative:

- Similar to A+, including higher risks of certain cancers, cardiovascular diseases, and autoimmune disorders.
- Increased sensitivity to H. pylori infection, leading to stomach issues.

Diseases Common in People with Type A Positive

Studies have shown that individuals with Type A or Type AB blood are at a higher risk for gastric cancer. Additionally, those with Type A, Type B, or Type AB blood may have an increased risk for pancreatic cancer. Other diseases more prevalent in people with Type A positive blood include spondyloarthropathy, diabetes mellitus, vasculitis, undifferentiated connective tissue disease, Behçet's disease, and rheumatoid arthritis.

BLOOD GROUP B POSITIVE

Biological Composition: Blood group B positive (B+) is characterized by the presence of the B antigen (D-galactose) on the surface of red blood cells. The B allele encodes a glycosyltransferase that produces this antigen. Blood group antigens are secondary gene products, with primary gene products being various glycosyltransferase enzymes that attach sugar molecules to the oligosaccharide chain.

Ethnicity & Population: Only 9% of the blood donor population has B-positive blood. It is more common among Americans of Asian descent (25%) and African Americans (20%).

Benefits:

- B-positive blood can be transfused to patients with B-positive and AB-positive blood.

- B-positive individuals can receive blood from all B and O types.
- Ideal donation types are whole blood and platelets.

Foods to Avoid:
- Corn, buckwheat, tomatoes, peanuts, sesame seeds, wheat, chicken, and certain fish.
- Inflammatory foods such as artichokes, avocado, pumpkin, radishes, tempeh, tofu, and certain grains like corn, wheat, barley, rye, and buckwheat.

Recommended Foods:
- Green vegetables, eggs, low-fat dairy, oats, milk products, animal protein, oat bran, paneer, fish, oatmeal, and quinoa.

Diseases More Common in People with Type B Positive:
- Higher risk of type 2 diabetes mellitus (T2DM).
- Rh incompatibility issues during pregnancy if the mother is Rh-negative and the fetus is Rh-positive.

BLOOD GROUP B NEGATIVE

Biological Composition: Blood group B negative (B-) has B antigens but lacks the Rh factor. The B allele encodes a glycosyltransferase that produces the B antigen (D-galactose).

Ethnicity & Population: Only 2% of the population has B-negative blood. This type is rare, making donations crucial to ensure blood products are available when needed.

Benefits:
- B-negative blood can be transfused to B and AB patients.
- Rare donations are vital for maintaining a steady blood supply.

Foods to Avoid:
- Corn, wheat, buckwheat, lentils, tomatoes, peanuts, sesame seeds, chicken, and certain fish.
- Inflammatory foods such as artichokes, avocado, pumpkin, radishes, tempeh, tofu, and certain grains like corn, wheat, barley, rye, and buckwheat.

Recommended Foods:
- Dark leafy greens like kale, broccoli, collard greens, mustard greens.
- Beets, sweet potatoes, carrots, and cauliflower.
- Fruits like pineapples, cranberries, grapes, plums, and papaya.

Diseases More Common in People with Type B Negative:
- Higher risk of cognitive and memory problems, including dementia, compared to those with Type O.
- Increased susceptibility to certain cancers and autoimmune diseases.

BLOOD GROUP AB POSITIVE

Biological Composition: Blood group AB positive (AB+) has both A and B antigens and the Rh factor on the red blood cells.

Ethnicity & Population: AB-positive is one of the rarer blood types, present in only 4% of the U.S. blood donor population.

Benefits:
- AB-positive is the "universal recipient" as patients can receive red blood cells from all blood types.

Foods to Avoid:
- Corn, wheat, buckwheat, lentils, tomatoes, peanuts, chicken, beef, smoked or cured meat, and sesame seeds.
- Avoid caffeine, alcohol, whole milk, coconut, bananas, mangoes, and black tea.

Recommended Foods:
- Dairy, tofu, lamb, fish, grains, fruit, and vegetables.
- Lean meat, fish, eggs, kale, lettuce, broccoli, onions, pumpkins, turnip, red peppers, okra, garlic, ginger, cherries, figs, plums, prunes, raspberries, cranberries, gooseberries, animal protein, fermented foods.

Diseases More Common in People with Type AB Positive:
- Higher risk for gastric and pancreatic cancer.
- Increased likelihood of cognitive and memory problems leading to dementia.

BLOOD GROUP AB NEGATIVE

Biological Composition: AB-negative is the rarest blood type, with only 1% of the U.S. blood donor population having this type. It is defined by the presence of both A and B antigens but lacks the Rh factor.

Ethnicity & Population: AB-negative is the rarest, making up only 1% of the population. B-negative is also rare, at 2%, followed by AB-positive at 4%.

Benefits:
- AB-negative is a universal plasma donor, meaning anyone can receive AB-negative plasma.
- AB-negative patients can receive any Rh-negative blood type.

Foods to Avoid:
- Corn, wheat, buckwheat, lentils, tomatoes, peanuts, chicken, beef, smoked or cured meat, and sesame seeds.
- Avoid caffeine, alcohol, whole milk, coconut, bananas, mangoes, and black tea.

Recommended Foods:
- Fruits, vegetables, whole grains, legumes, and nuts.
- Limited animal protein intake.

Diseases More Common in People with Type AB Negative:
- Higher risk of gastric and pancreatic cancer compared to blood type O.
- Vulnerability to exocrine pancreatic cancer, particularly in blood types B and AB.

BLOOD TYPE DIET RECOMMENDATIONS

Type O Blood:
- High-protein diet: lean meat, poultry, fish, vegetables.
- Light on grains, beans, and dairy.

- Supplements for digestive issues.

Type A Blood:
- Meat-free diet: fruits, vegetables, beans, legumes, whole grains.
- Prefer organic and fresh foods.

Type B Blood:
- Avoid corn, wheat, buckwheat, lentils, tomatoes, peanuts, sesame seeds.
- Recommended: green vegetables, eggs, certain meats, low-fat dairy.

Type AB Blood:
- Focus on tofu, seafood, dairy, green vegetables.
- Avoid caffeine, alcohol, smoked or cured meats.

BLOOD TYPE GENETICS

Blood types are determined by the combination of alleles inherited from parents. Blood types A and B are co-dominant, while blood type O is recessive. The possible combinations of alleles and their corresponding blood types are as follows:

- **AB:** Blood type AB
- **BB or BO:** Blood type B
- **AA or AO:** Blood type A
- **OO:** Blood type O

Blood Type Compatibility:

- **Type A:** Can receive type A and type O blood.
- **Type B:** Can receive type B and type O blood.
- **Type AB:** Can receive type A, B, AB, and O blood, making them universal recipients.
- **Type O:** Can donate to any blood type but can only receive type O blood, making them universal donors.

THE RELATIONSHIP OF BLOOD GROUP WITH DISEASE RISKS

The table below summarizes the associations between blood groups, antigens, and various diseases:

Disease	Risk Factor	Blood Group/Antigens
Sickle cell anemia	Increased adhesion	Adhesion Molecules
Hemolytic disease of the newborn	Antibodies to RhD	RhD
Chronic and autoimmune hemolytic anemias	Rh null	Rh, RhAG
Vascular disorders, venous and arterial thromboembolism, coronary heart disease, ischemic stroke, myocardial infarction	Reduced clearance of von Willebrand factor and FVIII	Groups A > AB > B
Dementia, cognitive impairment	Coagulation factors	Groups AB > B > A

Disease	Risk Factor	Blood Group/ Antigens
Plague, cholera, tuberculosis, mumps	Antigen profile	Group O
Smallpox, Pseudomonas aeruginosa	Antigen profile	Group A
Gonorrhea, tuberculosis, S. pneumoniae, E. coli, salmonella	Antigen profile	Group B
Smallpox, E. coli, salmonella	Antigen profile	Group AB
N. meningitides, H. influenza, C. albicans, S. pneumoniae, E. coli urinary tract infections, S. pyogenes, V. cholera	Antigen profile	Non-secretors
H. pylori	Strain-dependent	Group A: 95% non-O
Peptic ulcers, gastroduodenal disease	Secretor status, H. pylori strain	All non-secretors; Group O
Norovirus	Strain-dependent	Secretors; groups O, A
P. falciparum malaria	Receptor/antigen profile	Knops antigens; groups A, B
P. vivax malaria	Antigen profile	Duffy FY antigens
Cholera	Severity differs by antigen profile	Lewis' antigen; non-secretors; non-O groups

Disease	Risk Factor	Blood Group/ Antigens
Bacterial Meningitis (N. meningitides, H. influenza, S. pneumoniae)	Antigen profile	Non-secretors: A, AB, O blood groups
Cancer (tissue-specific)	Increased tumor antigens and ligands	A, B, H antigens lost; "A-like" antigens gained
Leukemia and Lymphoma	RBC membrane changes	A, B, H antigens lost
Non-Hodgkin's central nervous system lymphoma (primary and secondary)		Group O, B
Hodgkin's lymphoma		Group B
Acute lymphoblastic leukemia		Group O
Acute myeloid leukemia		Group A
Stomach Cancer	H. pylori strain	Group A
Pancreatic Cancer	H. pylori strain	Group B > AB > A
Von Hippel-Lindau and Neuroendocrine	Multiple tumors	Group O
Multiple Endocrine Neoplasia Type 1	Strongly associated	Group O
Colon/Rectum Cancer	Type 1 and 2 chains; Lewis's antigens	Secretors: "A-like" antigens expressed

Disease	Risk Factor	Blood Group/ Antigens
Hypertension	Three phenotypes differ	Group B > A > AB
Hyperlipidemia	Low-fat diet ineffective; intestinal ALP and apoB-48 vary by secretor status	LDL: Heterozygous MN; Group A, B; ALP/apoB-48: Group O and B secretors
Type 2 Diabetes	Rh group modifies	Group AB > B > A
Type 1 Diabetes	FUT2 gene locus	Non-secretors

The risk factors and blood groups or antigens associated with various diseases are based on the research presented in this review. Understanding these associations can help in the diagnosis, treatment, and prevention of diseases based on an individual's blood type.

PART TWO
ALLERGIES

Chapter Four
OVERVIEW OF FOOD ALLERGIES

Food allergies are an increasingly common health concern affecting individuals of all ages. These allergic reactions occur when the body's immune system incorrectly identifies certain proteins in food as harmful invaders and mounts a defensive response. This response can manifest in various ways, from mild skin irritations to severe, life-threatening anaphylactic reactions. With the growing prevalence of food allergies, it is essential to understand their impact on individuals' daily lives, the healthcare system, and society as a whole. This book aims to provide a comprehensive resource for understanding and managing food allergies effectively.

DEFINITION AND IMPORTANCE

Food allergies are a specific type of adverse immune response to food proteins. Unlike food intolerances, which are typically caused by digestive issues and do not involve the immune system, food allergies can trigger a range of symptoms that can escalate to life-threatening levels. The importance of understanding food allergies lies in the potential severity of these reactions and the significant lifestyle adjustments required to avoid allergens. Effective management and awareness can prevent serious health incidents and improve the quality of life for those affected.

BRIEF HISTORY AND RISE IN PREVALENCE

The recognition of food allergies as a medical condition is relatively recent, although adverse reactions to food have been documented for centuries. Historical references to food-induced ailments can be found in ancient texts, but it was not until the 20th century that food allergies began to be studied and understood scientifically. Over the past few decades, there has been a marked increase in the incidence of food allergies, particularly in developed countries. Factors such as changes in dietary habits, environmental exposures, and genetic predispositions are believed to contribute to this rise. Understanding the historical context and trends in prevalence helps frame the current challenges and directions for research and public health initiatives.

PURPOSE AND SCOPE OF THE BOOK

This book is designed to serve as a comprehensive guide for anyone affected by or interested in food allergies. The purpose is to provide clear, evidence-based information on the nature of food allergies, methods for diagnosis, strategies for management, and recent advances in treatment. The scope of the book encompasses the scientific, medical, and practical aspects of living with food allergies. It is intended to be a valuable resource for individuals with food allergies, their families, healthcare professionals, educators, and anyone seeking to increase their knowledge and support those with this condition.

WHAT ARE FOOD ALLERGIES?

UNDERSTANDING THE BASICS

Food allergies are complex conditions resulting from the immune system's inappropriate response to certain food proteins. When a person with a food allergy consumes an allergenic food, their immune system perceives these proteins as threats and activates a defense mechanism. This section explores the fundamental concepts of food allergies, including the biological processes involved and the various ways these allergies can present.

IMMUNE SYSTEM RESPONSE

The immune system plays a crucial role in defending the body against harmful substances. In the case of food allergies, the immune response is triggered by specific proteins in food, known as allergens. The primary immune response involves the production of immunoglobulin E (IgE) antibodies. When an allergic individual ingests the offending food, these IgE antibodies bind to mast cells and basophils, which are types of white blood cells. Upon subsequent exposure to the allergen, these cells release histamine and other chemicals, leading to the symptoms of an allergic reaction. These symptoms can range from mild (such as itching and hives) to severe (such as difficulty breathing and anaphylaxis).

COMMON ALLERGENS

Food allergies can be triggered by a variety of foods, but some allergens are more common than others. The "big eight" allergens responsible for the majority of allergic reactions include:

- Peanuts

- Tree nuts (e.g., almonds, walnuts, cashews)
- Milk
- Eggs
- Wheat
- Soy
- Fish
- Shellfish

These foods are the most common culprits, but it is essential to recognize that allergies can develop to any food protein. Additionally, the prevalence of specific food allergens can vary across different regions and cultures. Understanding the common allergens helps in identifying and managing food allergies more effectively.

DIFFERENCE BETWEEN FOOD ALLERGIES AND FOOD INTOLERANCES

It is crucial to distinguish between food allergies and food intolerances, as they involve different mechanisms and management strategies. Food allergies are immune-mediated responses that can lead to severe, potentially life-threatening reactions. In contrast, food intolerances are generally related to the digestive system's inability to process certain substances in food. For example, lactose intolerance occurs due to a deficiency in lactase, the enzyme needed to digest lactose in dairy products. While food intolerances can cause discomfort and digestive issues, they do not involve the immune system and are not life-threatening. Understanding the differences between these conditions is essential for accurate diagnosis and appropriate management.

Chapter Five

DIAGNOSING FOOD ALLERGIES

SYMPTOMS AND SIGNS

Immediate vs. Delayed Reactions

Food allergies can manifest in various ways, and understanding the timing of reactions is crucial for accurate diagnosis. Reactions can be classified into two categories: immediate and delayed.

- **Immediate Reactions:** These typically occur within minutes to a couple of hours after consuming the allergen. Immediate reactions are often IgE-mediated and can range from mild to severe. Symptoms can escalate rapidly, making it essential to recognize and respond promptly.

- **Delayed Reactions:** These can take several hours to days to appear and are usually not IgE-mediated. Delayed reactions often involve the gastrointestinal tract and can be more challenging to diagnose due to their gradual onset.

COMMON SYMPTOMS IN CHILDREN AND ADULTS

Food allergy symptoms can vary widely among individuals and can affect multiple organ systems. Recognizing these symptoms is key to identifying potential food allergies.

- **In Children:**
    - **Skin:** Hives, eczema, itching, and swelling
    - **Gastrointestinal:** Vomiting, diarrhea, abdominal pain
    - **Respiratory:** Sneezing, nasal congestion, coughing, wheezing
    - **Anaphylaxis:** Severe, life-threatening reaction involving multiple systems, requiring immediate medical attention
- **In Adults:**
    - **Skin:** Hives, angioedema (swelling beneath the skin), itching
    - **Gastrointestinal:** Nausea, vomiting, abdominal cramps, diarrhea
    - **Respiratory:** Shortness of breath, wheezing, throat tightness
    - **Cardiovascular:** Drop in blood pressure, dizziness, fainting
    - **Anaphylaxis:** As in children, adults can also experience severe anaphylactic reactions

Understanding these symptoms helps in early recognition and intervention, reducing the risk of severe outcomes.

DIAGNOSTIC METHODS

Accurate diagnosis of food allergies involves a combination of medical history, clinical evaluation, and specific diagnostic tests. The following methods are commonly used by healthcare professionals:

Skin Prick Tests

Skin prick testing is a widely used diagnostic tool for identifying IgE-mediated food allergies. During the test:

- **Procedure:** Small amounts of allergen extracts are applied to the skin, typically on the forearm or back. The skin is then pricked or scratched to introduce the allergen into the outer layer.

- **Reaction:** If the individual is allergic to the tested substance, a localized reaction (red, itchy bump) will appear within 15-20 minutes.

- **Interpretation:** The size of the wheal (raised bump) and surrounding redness are measured to assess the degree of sensitivity to the allergen.

Skin prick tests are quick and minimally invasive, providing valuable information about potential allergens.

Blood Tests

Blood tests, such as the specific IgE test (formerly known as RAST or ImmunoCAP), measure the levels of IgE antibodies to specific foods in the blood.

- **Procedure:** A blood sample is drawn and analyzed in a laboratory to detect IgE antibodies specific to various allergens.

- **Advantages:** Blood tests are useful when skin prick testing is not feasible, such as in individuals with extensive eczema or those who cannot discontinue antihistamines.

- **Interpretation:** Elevated IgE levels indicate a potential allergy, but results must be interpreted in conjunction with clinical history and other tests.

Blood tests provide an alternative method for diagnosing food allergies, particularly in complex cases.

Oral Food Challenges

The oral food challenge is considered the gold standard for diagnosing food allergies. It involves the supervised ingestion of the suspected allergen in a controlled setting.

- **Procedure:** The individual consumes gradually increasing amounts of the suspected allergen under medical supervision. The challenge is typically conducted in a hospital or specialized clinic equipped to manage allergic reactions.

- **Phases:** The challenge starts with small doses and progresses to larger amounts, with careful monitoring for any signs of a reaction.

- **Interpretation:** A positive reaction confirms the allergy, while the absence of symptoms indicates tolerance to the food.

Oral food challenges are definitive but require careful planning and medical oversight due to the risk of severe reactions.

Elimination Diets

Elimination diets involve removing suspected allergens from the diet for a period and then gradually reintroducing them to observe for reactions.

- **Procedure:** The individual eliminates one or more suspected allergens from their diet for a few weeks. If symptoms improve, the allergen is gradually reintroduced under medical guidance.

- **Monitoring:** A detailed food diary is kept to track symptoms and any changes in health.

- **Interpretation:** Improvement during the elimination phase followed by a recurrence of symptoms upon reintroduction suggests an allergy to the food.

Elimination diets are useful for identifying food allergies and intolerances, particularly when other diagnostic tests are inconclusive.

Diagnosing food allergies requires a comprehensive approach that includes recognizing symptoms, understanding the timing of reactions, and utilizing various diagnostic methods. Accurate diagnosis is essential for effective management and improving the quality of life for individuals with food allergies. Through a combination of skin prick tests, blood tests, oral food challenges, and elimination diets, healthcare professionals can identify specific allergens and develop tailored management plans.

Chapter Six

MANAGING FOOD ALLERGIES

Managing food allergies effectively requires a comprehensive approach that encompasses avoidance strategies, emergency preparedness, and lifestyle adjustments. This chapter delves into practical measures to help individuals with food allergies navigate daily life safely and confidently.

AVOIDANCE STRATEGIES

Reading Food Labels

One of the most crucial skills for managing food allergies is the ability to read and interpret food labels accurately. Manufacturers are required by law in many countries to list common allergens on food packaging, but it's important to be vigilant and thorough.

- **Understanding Labels:** Learn to recognize alternative names for allergens (e.g., casein for milk, albumin for eggs).

Familiarize yourself with terms that indicate potential cross-contamination, such as "may contain" or "processed in a facility that also processes."

- **Ingredient Lists:** Carefully read the ingredient list on packaged foods. Be aware that ingredients can change, so even familiar products should be checked regularly.

- **Allergen Statements:** Pay attention to allergen statements often found at the end of the ingredient list. These statements highlight the presence of major allergens and potential cross-contamination risks.

Being diligent in reading food labels can significantly reduce the risk of accidental exposure to allergens.

Cross-Contamination Prevention

Cross-contamination occurs when an allergen unintentionally comes into contact with allergen-free foods, posing a significant risk to individuals with food allergies. Implementing strategies to prevent cross-contamination is vital.

- **Kitchen Practices:** Designate separate utensils, cutting boards, and cooking surfaces for allergen-free foods. Clean all equipment and surfaces thoroughly with soap and water after contact with allergens.

- **Storage:** Store allergen-containing foods separately from allergen-free foods. Use clearly labeled, airtight containers to avoid accidental mixing.

- **Cooking:** Prepare allergen-free foods first to minimize the risk of cross-contamination. Avoid using shared oils or fryers for allergen-containing and allergen-free foods.

By adopting strict cross-contamination prevention measures, you can create a safer environment for those with food allergies.

Safe Eating Out Practices

Eating out can be challenging for individuals with food allergies, but with careful planning and communication, it is possible to enjoy dining out safely.

- **Research Restaurants:** Choose restaurants that are known for their allergy-friendly practices. Look for places that provide detailed menu information and are willing to accommodate special dietary needs.

- **Communicate Clearly:** Inform the restaurant staff about your food allergies as soon as you arrive. Clearly explain the severity of the allergy and the need to avoid cross-contamination.

- **Ask Questions:** Inquire about ingredients, preparation methods, and potential cross-contact risks. Don't hesitate to ask the chef or manager for detailed information.

- **Be Prepared:** Carry a chef card or allergy alert card that lists your allergens and necessary precautions. Always have your emergency medications, such as epinephrine auto-injectors, readily available.

Taking these steps can help ensure a safer dining experience and reduce the risk of allergic reactions when eating out.

EMERGENCY PLANS

Recognizing Anaphylaxis

Anaphylaxis is a severe, life-threatening allergic reaction that requires immediate medical attention. Being able to recognize the signs and symptoms of anaphylaxis is crucial.

- **Symptoms of Anaphylaxis:** Symptoms can include difficulty breathing, swelling of the throat or tongue, hives, rapid heartbeat, dizziness, and a drop in blood pressure. Gastrointestinal symptoms like nausea and vomiting can also occur.

- **Action Steps:** If you suspect anaphylaxis, administer epinephrine immediately and call emergency services. Do not wait to see if symptoms improve, as prompt treatment is essential.

Knowing how to recognize anaphylaxis can save lives by ensuring swift and appropriate action is taken.

Using Epinephrine Auto-Injectors

Epinephrine is the first-line treatment for anaphylaxis. Epinephrine auto-injectors are designed for quick and easy administration in emergency situations.

- **How to Use:** Familiarize yourself with the specific auto-injector prescribed. Common brands include EpiPen, Auvi-Q, and others. Practice using a trainer device to build confidence.

- **Administration Steps:** Remove the safety cap, position the injector against the outer thigh, and press firmly until you hear a click. Hold in place for the recommended time (usually 3-10 seconds). Seek immediate medical attention after administering epinephrine.

- **Storage:** Store auto-injectors at room temperature, away from direct sunlight and extreme temperatures. Check the expiration date regularly and replace as needed.

Understanding how to use epinephrine auto-injectors correctly can make a critical difference during an allergic emergency.

Creating an Emergency Action Plan

An emergency action plan is a detailed guide outlining the steps to take in the event of an allergic reaction. It should be shared with family members, caregivers, teachers, and anyone involved in the individual's care.

- **Components of the Plan:**
    - **Allergen Information:** List the specific allergens and common sources of exposure.
    - **Symptoms:** Describe the signs and symptoms of mild, moderate, and severe reactions.
    - **Response Steps:** Outline the steps to take for different levels of reactions, including when to administer epinephrine and call emergency services.
    - **Contact Information:** Provide emergency contact numbers for family members, healthcare providers, and emergency services.
- **Distribution:** Ensure that copies of the emergency action plan are readily available at home, school, work, and any other relevant locations. Review and update the plan regularly.

Creating and maintaining an emergency action plan ensures that everyone involved is prepared to respond effectively in the event of an allergic reaction, enhancing safety and peace of mind.

By implementing these avoidance strategies and emergency plans, individuals with food allergies can navigate their daily lives more safely and confidently, reducing the risk of accidental exposure and ensuring prompt, effective response in case of an allergic reaction.

NUTRITIONAL CONSIDERATIONS

Managing food allergies requires careful attention to nutritional needs to ensure a balanced diet while avoiding allergens. This chapter focuses on strategies for maintaining a healthy diet, alternative sources of essential nutrients, and the use of safe substitutes and supplements.

MAINTAINING A BALANCED DIET

Alternative Sources of Nutrients

For individuals with food allergies, finding alternative sources of essential nutrients is crucial to maintaining overall health. Here are some common allergens and their alternative nutrient sources:

- **Milk Allergy:**
    - **Calcium:** Leafy green vegetables (e.g., kale, broccoli), fortified non-dairy milk (e.g., almond, soy, oat), fortified orange juice.
    - **Vitamin D:** Sunlight exposure, fortified non-dairy milk, supplements.
    - **Protein:** Legumes (e.g., beans, lentils), tofu, quinoa, nuts, and seeds (if not allergic).

- **Egg Allergy:**
  - **Protein:** Meat, poultry, fish, legumes, nuts, and seeds (if not allergic), tofu, tempeh.
  - **Vitamin B12:** Fortified cereals, nutritional yeast, supplements.

- **Peanut and Tree Nut Allergy:**
  - **Healthy Fats:** Seeds (e.g., chia, flax, sunflower), avocado, olive oil.
  - **Protein:** Seeds, legumes, lean meats, fish, dairy products (if not allergic).

- **Wheat Allergy:**
  - **Carbohydrates:** Rice, quinoa, oats (if gluten-free), corn, potatoes, sweet potatoes.
  - **Fiber:** Fruits, vegetables, legumes, gluten-free whole grains.

- **Soy Allergy:**
  - **Protein:** Meat, poultry, fish, eggs (if not allergic), dairy products (if not allergic), legumes.
  - **Iron and Zinc:** Meat, poultry, fish, fortified cereals.

Ensuring a diverse diet with a variety of foods can help meet nutritional needs while avoiding allergens.

Meal Planning and Preparation

Effective meal planning and preparation are essential for managing food allergies and ensuring a balanced diet. Here are some tips:

- **Weekly Meal Planning:** Plan meals and snacks for the

week, focusing on incorporating a variety of safe foods. Use a meal planning template to organize your shopping list and ensure you have all necessary ingredients.

- **Batch Cooking:** Prepare larger portions of allergen-free meals and freeze individual servings for convenient, quick meals throughout the week.

- **Safe Kitchen Practices:** Maintain a clean and organized kitchen to prevent cross-contamination. Use separate utensils and cooking equipment for allergen-free foods.

- **Diverse Recipes:** Experiment with new recipes and ingredients to keep meals interesting and nutritious. Resources such as allergy-friendly cookbooks and websites can provide inspiration.

By planning and preparing meals in advance, individuals with food allergies can ensure they have safe and nutritious options readily available.

SUPPLEMENTS AND SUBSTITUTES

Safe Substitutes for Common Allergens

Finding safe substitutes for common allergens is crucial for maintaining a balanced diet without compromising on taste and nutrition. Here are some alternatives:

- **Milk Alternatives:**
  - **Non-Dairy Milk:** Almond milk, soy milk, oat milk, rice milk, coconut milk (ensure they are fortified with calcium and vitamin D).
  - **Non-Dairy Yogurt and Cheese:** Products made from soy, almond, coconut, and other plant-based ingredients.

- **Egg Alternatives:**
  - **Baking:** Use applesauce, mashed bananas, yogurt, or commercial egg replacers.
  - **Cooking:** Use tofu or chickpea flour for egg-free scrambles and omelets.

- **Peanut and Tree Nut Alternatives:**
  - **Seed Butters:** Sunflower seed butter, pumpkin seed butter.
  - **Other Spreads:** Soy nut butter (if not allergic), tahini (sesame seed butter).

- **Wheat Alternatives:**
  - **Flours:** Rice flour, almond flour (if not allergic), coconut flour, chickpea flour, gluten-free flour blends.
  - **Grains:** Quinoa, rice, millet, buckwheat, sorghum.

- **Soy Alternatives:**
  - **Soy Sauce:** Use coconut aminos or tamari (gluten-free soy sauce).
  - **Protein:** Meat, poultry, fish, legumes, eggs (if not allergic).

By incorporating these substitutes, individuals with food allergies can enjoy a variety of meals without sacrificing nutritional value.

Nutritional Supplements

In some cases, it may be challenging to meet all nutritional needs through diet alone, especially when multiple food allergies are present. Nutritional supplements can help fill these gaps, but it is essential to choose them carefully:

- **Multivitamins:** Consider a comprehensive multivitamin that includes essential vitamins and minerals, ensuring it is free from allergens.

- **Calcium and Vitamin D:** For those with milk allergies, calcium and vitamin D supplements can help maintain bone health.

- **Omega-3 Fatty Acids:** Fish oil supplements or algae-based omega-3 supplements can provide essential fatty acids, especially for those allergic to fish or nuts.

- **Probiotics:** Probiotics can support gut health, particularly for individuals with gastrointestinal symptoms related to food allergies.

Before starting any supplements, consult with a healthcare provider to ensure they are necessary and appropriate for your specific needs.

Chapter Seven
SPECIAL POPULATIONS

Food allergies affect people of all ages, but certain populations face unique challenges and considerations. This chapter focuses on the specific needs and management strategies for children with food allergies and adults who develop food allergies later in life.

CHILDREN WITH FOOD ALLERGIES

Infant and Toddler Considerations

Managing food allergies in infants and toddlers requires special attention due to their developmental stage and nutritional needs.

- **Breastfeeding and Formula:** Breastfeeding is recommended for infants with a family history of allergies, as it may provide protective benefits against the development of food allergies (Greer et al., 2008). For infants who require formula, hypoallergenic formulas such as extensively hydrolyzed or amino acid-based formulas are recommended for those at high risk of allergies or who have already developed allergies (Sicherer & Sampson, 2014).

- **Introduction of Solid Foods:** Introducing solid foods should be done carefully and under the guidance of a healthcare provider. Recent guidelines suggest introducing allergenic foods such as peanuts and eggs early (around 4-6 months of age) to potentially reduce the risk of developing food allergies (Du Toit et al., 2015).

- **Monitoring for Reactions:** Parents and caregivers should be vigilant in monitoring for signs of allergic reactions when introducing new foods. Symptoms can include hives, vomiting, diarrhea, and in severe cases, difficulty breathing.

- **Emergency Preparedness:** It's essential to have an emergency action plan in place, including access to antihistamines and epinephrine auto-injectors, in case of severe allergic reactions (anaphylaxis).

School-age Children

School-age children with food allergies face additional challenges as they navigate social environments and increased independence.

- **Education and Awareness:** Educating children about their food allergies is crucial. They should understand what foods to avoid, how to read food labels, and the importance of not sharing food with others. Schools should be informed about the child's allergies, and staff should be trained in recognizing and managing allergic reactions.

- **School Policies:** Schools should have policies in place to manage food allergies, including allergen-free zones, safe food handling practices, and access to emergency medications (Turner et al., 2015). An individualized healthcare plan (IHP) and an emergency action plan should be developed in collaboration with the school nurse and healthcare provider.

- **Social Inclusion:** It's important to ensure that children with food allergies do not feel excluded. Participation in school activities, parties, and field trips should be planned with their safety in mind, with alternative food options provided as needed.

- **Peer Support:** Encouraging a supportive peer environment can help children with food allergies feel more comfortable and confident. Education programs for classmates can foster understanding and empathy.

ADULTS WITH NEW ONSET FOOD ALLERGIES

Adjusting to New Diagnoses

Adults who develop food allergies later in life may face unique challenges in adjusting to their new diagnoses.

- **Emotional Impact:** Receiving a food allergy diagnosis as an adult can be stressful and overwhelming. Adults may experience anxiety, frustration, and a sense of loss, particularly if the allergy affects foods that were previously enjoyed. Support groups and counseling can provide emotional support and coping strategies (LeBovidge et al., 2008).

- **Lifestyle Adjustments:** Adults will need to make significant lifestyle changes to manage their food allergies safely. This includes learning to read food labels, avoiding cross-contamination, and being vigilant about food choices when dining out.

- **Dietary Management:** Adjusting to a restricted diet requires careful planning to ensure nutritional needs are met. Working with a dietitian can help adults develop a balanced diet that avoids allergens while providing essential nutrients.

- **Workplace Considerations:** Managing food allergies in the workplace involves informing colleagues and employers about the allergy and implementing strategies to prevent exposure. This may include designated eating areas, safe food handling practices, and ensuring emergency medications are readily available.

Lifestyle Changes

Adapting to food allergies as an adult involves several lifestyle changes to ensure safety and well-being.

- **Home Environment:** Adults will need to make their home environment safe by eliminating allergens from their pantry, thoroughly cleaning surfaces, and educating family members about cross-contamination risks.

- **Travel:** Traveling with food allergies requires additional planning. Researching restaurants, bringing safe snacks, and carrying a chef card that explains the allergy in the local language can help manage risks when traveling abroad.

- **Social Activities:** Participating in social activities, such as dining out or attending events, requires careful communication with hosts and restaurant staff to ensure safe food options. Adults may need to bring their own food to events to avoid accidental exposure.

- **Continued Education:** Staying informed about food allergies, new treatments, and management strategies is important. Regular follow-ups with an allergist can help manage the allergy effectively and stay updated on any new developments.

By addressing the unique needs and challenges of children and adults with food allergies, this chapter aims to provide practical guidance and support for managing food allergies across different life stages.

RECIPES AND MEAL PLANS

Creating delicious and nutritious meals that are safe for individuals with food allergies can be a rewarding challenge. This chapter provides a variety of allergy-friendly recipes for all meals and occasions, along with sample meal plans and tips for adapting family meals to accommodate various allergens.

ALLERGY-FRIENDLY RECIPES

Breakfast, Lunch, Dinner, and Snacks

Finding safe and enjoyable recipes for every meal of the day is crucial for maintaining a balanced diet and avoiding allergens. Here are some versatile, allergy-friendly recipes to get you started:

Breakfast:

- **Gluten-Free Oatmeal:**
    - Ingredients: Gluten-free oats, almond milk (or other non-dairy milk), fresh berries, chia seeds, maple syrup.
    - Preparation: Cook oats in almond milk according to package instructions. Top with fresh berries, chia seeds, and a drizzle of maple syrup.

- **Egg-Free Pancakes:**
    - Ingredients: Gluten-free flour, baking powder, unsweetened applesauce, non-dairy milk, vanilla extract.

- Preparation: Mix dry ingredients. In a separate bowl, combine wet ingredients. Combine both mixtures and cook pancakes on a non-stick skillet.

Lunch:

- **Turkey and Avocado Wrap:**
  - Ingredients: Gluten-free tortilla, sliced turkey breast, avocado, lettuce, tomato, dairy-free mayo.
  - Preparation: Spread dairy-free mayo on the tortilla. Layer with turkey, avocado slices, lettuce, and tomato. Roll up and serve.

- **Quinoa Salad:**
  - Ingredients: Cooked quinoa, chickpeas, cucumber, cherry tomatoes, red onion, olive oil, lemon juice, fresh herbs.
  - Preparation: Combine quinoa, chickpeas, and chopped vegetables in a bowl. Dress with olive oil, lemon juice, and herbs. Toss and serve chilled.

Dinner:

- **Chicken and Vegetable Stir-Fry:**
  - Ingredients: Chicken breast, bell peppers, broccoli, snap peas, gluten-free soy sauce (or coconut aminos), garlic, ginger, olive oil.
  - Preparation: Sauté garlic and ginger in olive oil. Add chicken and cook until browned. Add vegetables and stir-fry until tender. Stir in gluten-free soy sauce and serve over rice.

- **Sweet Potato and Black Bean Tacos:**
  - Ingredients: Sweet potatoes, black beans, corn tortillas, avocado, salsa, lime, cilantro.
  - Preparation: Roast sweet potato cubes until tender. Warm tortillas and fill with sweet potatoes, black beans, avocado slices, and salsa. Garnish with lime and cilantro.

Snacks:

- **Hummus and Veggie Sticks:**
  - Ingredients: Homemade or store-bought hummus, carrot sticks, celery sticks, cucumber slices.
  - Preparation: Serve hummus with an assortment of fresh vegetable sticks.

- **Energy Bites:**
  - Ingredients: Rolled oats, sunflower seed butter, honey, flax seeds, dried cranberries.
  - Preparation: Mix all ingredients in a bowl. Form into bite-sized balls and refrigerate until firm.

Desserts and Special Occasions

Enjoying dessert and celebrating special occasions is possible with these allergy-friendly recipes that everyone will love.

- **Dairy-Free Chocolate Cake:**
  - Ingredients: Gluten-free flour, cocoa powder, baking soda, coconut oil, unsweetened applesauce, almond milk, vanilla extract.

- Preparation: Mix dry ingredients. In a separate bowl, combine wet ingredients. Combine both mixtures and bake in a preheated oven until a toothpick comes out clean. Frost with dairy-free chocolate frosting.

- **Fruit Sorbet:**
  - Ingredients: Fresh fruit (e.g., mango, berries), lemon juice, honey.
  - Preparation: Blend fruit, lemon juice, and honey until smooth. Freeze in an ice cream maker or a shallow dish, stirring occasionally until firm.

- **Allergen-Free Cupcakes:**
  - Ingredients: Gluten-free flour, baking powder, sugar, non-dairy milk, vegetable oil, vanilla extract.
  - Preparation: Mix dry ingredients. In a separate bowl, combine wet ingredients. Combine both mixtures and bake in a preheated oven. Frost with allergen-free frosting.

SAMPLE MEAL PLANS

Weekly Meal Plans for Various Allergens

Creating weekly meal plans can simplify grocery shopping and ensure a balanced diet. Here are sample meal plans for different allergens:

Gluten-Free Meal Plan:
- **Monday:**
  - Breakfast: Smoothie bowl with almond milk, bananas, and berries.

- Lunch: Quinoa salad with chickpeas and vegetables.
- Dinner: Grilled salmon with roasted vegetables.

- **Tuesday:**
    - Breakfast: Egg-free pancakes with maple syrup.
    - Lunch: Turkey and avocado wrap.
    - Dinner: Chicken and vegetable stir-fry.

Dairy-Free Meal Plan:
- **Wednesday:**
    - Breakfast: Gluten-free oatmeal with almond milk.
    - Lunch: Lentil soup with a side salad.
    - Dinner: Sweet potato and black bean tacos.

- **Thursday:**
    - Breakfast: Smoothie with coconut milk, spinach, and pineapple.
    - Lunch: Hummus and veggie sticks.
    - Dinner: Baked chicken with quinoa and steamed broccoli.

Nut-Free Meal Plan:
- **Friday:**
    - Breakfast: Sunflower seed butter on gluten-free toast.
    - Lunch: Quinoa salad with chickpeas and vegetables.
    - Dinner: Beef stir-fry with rice and vegetables.

- **Saturday:**
  - Breakfast: Egg-free pancakes with sunflower seed butter.
  - Lunch: Turkey and avocado wrap.
  - Dinner: Grilled shrimp with roasted sweet potatoes.

TIPS FOR ADAPTING FAMILY MEALS

Adapting family meals to accommodate food allergies ensures that everyone can enjoy the same dishes safely. Here are some tips:

- **Identify Safe Substitutes:** Use allergy-friendly substitutes that do not compromise the taste or texture of meals. For example, use coconut milk instead of cow's milk, or flaxseed meal as an egg replacer.

- **Modify Recipes:** Adjust favorite family recipes by replacing allergenic ingredients with safe alternatives. Test these modifications to ensure they still meet your family's taste preferences.

- **Cook in Batches:** Prepare allergen-free versions of family favorites in larger quantities and freeze portions for future meals.

- **Separate Cooking Areas:** Use separate cooking utensils and surfaces for preparing allergen-free foods to avoid cross-contamination.

- **Involve the Family:** Educate family members about food allergies and involve them in meal planning and preparation. This can create a supportive environment and increase understanding of dietary restrictions.

By incorporating these recipes, meal plans, and adaptation tips, individuals and families can enjoy a variety of delicious, allergy-friendly meals while maintaining a balanced and nutritious diet. This chapter aims to make cooking and eating with food allergies a delightful and stress-free experience.

ADDITIONAL NOTES

- **Cherries, peaches, plums, apricots, almonds, and cashews** are rich in vitamins, minerals, and antioxidants, contributing to overall health and wellness.

- **Portion control** is essential to avoid potential side effects and maximize benefits.

- **Consult healthcare professionals** if you have underlying health conditions or food sensitivities before making significant changes to your diet.

PART THREE
Foods, Benefits & Daily Recommendations

Chapter Eight
TYPES OF FOODS, POSSIBLE SIDE EFFECTS, AND QUANTITIES RECOMMENDED

CHERRIES

Benefits:

- Effective in reducing the risks of gout by lowering uric acid levels.
- Help reduce inflammation, oxidative stress, muscle soreness, and lower blood pressure.
- Contain antioxidants that protect against chronic diseases such as cardiovascular disease, diabetes, cancer, and obesity.
- Bioactive compounds reduce symptoms of arthritis and osteoarthritis.
- Vitamin C content helps eliminate uric acid, beneficial for gout patients.

Possible Side Effects:
- May trigger bowel irritation in individuals with irritable bowel syndrome due to sugars like fructose and sorbitol.
- Some individuals may have latex and fruit syndrome, a cross-reactivity with latex allergy.

Quantity Recommendations:
- Recommended daily intake is about 20-25 cherries.
- Consuming whole cherries, including the peel, is essential for maximizing health benefits.
- Eating cherries or drinking cherry juice can reduce inflammation and pain.

PEACHES

Benefits:
- High water and fiber content aids in digestion and reduces constipation risk.
- Rich in vitamin C, which boosts immunity and fights colds.
- Beneficial for skin health due to antioxidants, beta-carotene, and vitamins A and C.
- Helps maintain blood pressure due to high potassium content.

Possible Side Effects:
- May cause allergies in hypersensitive individuals.
- Excessive consumption may lead to weight gain and hyperkalemia due to high potassium levels.

Quantity Recommendations:
- Consume one large or two small peaches daily.
- Peaches are low in calories and high in fiber, making them suitable for regular consumption.

PLUMS

Benefits:
- Rich in antioxidants and anti-inflammatory compounds.
- High fiber and sorbitol content promote gut health and relieve constipation.
- Good source of vitamins and minerals like calcium and magnesium, essential for bone health.
- Potassium helps control high blood pressure.

Possible Side Effects:
- High intake may cause digestive distress and kidney stones.
- Possible allergic reactions in individuals sensitive to plants in the Rosaceae family.

Quantity Recommendations:
- Recommended daily intake is 1-2 plums.
- Dried plums (prunes) can be consumed to treat constipation.

APRICOTS

Benefits:
- Rich in vitamins, minerals, and beneficial compounds like vitamin C, carotenoids, potassium, and polyphenols.

- Vitamin C supports immune function and promotes iron absorption.
- Carotenoids promote eye health.
- Fiber content aids in weight maintenance and blood sugar regulation.

Possible Side Effects:
- May cause allergies in sensitive individuals.
- Dried apricots containing sulphites may trigger allergies.

Quantity Recommendations:
- Consume 3-5 portions of apricots daily as part of a balanced diet.
- Avoid excessive intake to prevent potential health issues.

ALMONDS

Benefits:
- High in essential nutrients like healthy monounsaturated fats, fiber, protein, vitamins, and minerals.
- Rich in magnesium, which helps control blood sugar, lower blood pressure, and aid muscle recovery.
- Fiber and protein content can reduce hunger and help manage weight.

Possible Side Effects:
- Excessive consumption can interfere with micronutrient absorption due to phytic acid.
- May cause obstipation and allergies.

Quantity Recommendations:
- Recommended daily intake is about 15 almonds.
- Include almonds as part of a plant-based diet for cardiovascular benefits.

CASHEWS

Benefits:
- Fiber-rich, low in sugars, and contain healthy fats.
- Rich in antioxidants, improving immune function and preventing chronic diseases.
- Contain vitamin C and zinc, essential for strengthening the immune system.

Possible Side Effects:
- High oxalate content may increase kidney stone risk in susceptible individuals.
- Potential interactions with medications like antibiotics and blood pressure drugs.
- Allergies to cashews are increasing and can cause severe reactions.

Quantity Recommendations:
- Daily intake should be 20-30 grams of cashews.
- Include a variety of nuts in the diet to maximize health benefits.

PEACHES (REPEATED ENTRY)

Possible Side Effects:

- May cause allergies in hypersensitive individuals.
- Excessive consumption may lead to weight gain and hyperkalemia due to high potassium levels.

Quantity Recommendations:

- Consume one large or two small peaches daily.
- Peaches are low in calories and high in fiber, making them suitable for regular consumption.
- 8. Plums (Repeated Entry)

Benefits:

- Rich in antioxidants and anti-inflammatory compounds.
- High fiber and sorbitol content promote gut health and relieve constipation.
- Good source of vitamins and minerals like calcium and magnesium, essential for bone health.
- Potassium helps control high blood pressure.

Possible Side Effects:

- High intake may cause digestive distress and kidney stones.
- Possible allergic reactions in individuals sensitive to plants in the Rosaceae family.

Quantity Recommendations:

- Recommended daily intake is 1-2 plums.
- Dried plums (prunes) can be consumed to treat constipation.

ADDITIONAL NOTES

- **Cherries, peaches, plums, apricots, almonds, and cashews** are rich in vitamins, minerals, and antioxidants, contributing to overall health and wellness.

- **Portion control** is essential to avoid potential side effects and maximize benefits.

- **Consult healthcare professionals** if you have underlying health conditions or food sensitivities before making significant changes to your diet.Top of Form

PISTACHIO

Benefits:

- Effective in reducing the risk of cardiovascular diseases by lowering total and LDL cholesterol and increasing HDL cholesterol due to monosaturated fatty acids like oleic acid.
- Rich in phytonutrients such as carotene and polyphenolic compounds, which help remove toxic free radicals.
- High in vitamin E, which helps maintain the integrity of mucosa and skin and acts as a free radical scavenger.
- Can help control blood pressure due to high potassium content.
- Support glycemic control due to a low glycemic index, beneficial for individuals with prediabetes or insulin resistance.
- May aid in weight loss as a portion-controlled snack rich in fiber and protein.

Possible Side Effects:

- Individuals with kidney stones should consult healthcare practitioners before consuming pistachios, as they can cause kidney damage and diarrhea due to high protein content.
- Allergic reactions are common and typically develop early in life.

Quantity Recommendations:

- Consume no more than 80 grams of pistachios daily due to their high calorie content.
- The American Heart Association recommends four servings of unsalted and unoiled nuts weekly, with a serving size of about 1/3 cup of pistachios.

WALNUT

Benefits:

- Rich in antioxidants such as vitamin E, melatonin, and polyphenols, which protect cells from free radical damage.
- May help manage type 2 diabetes, cardiovascular diseases, and some cancers due to high fiber and polyunsaturated fat content.
- Regular consumption can prevent dementia and improve memory due to high levels of alfa-linolenic acid and polyphenols.
- Potassium content helps decrease high blood pressure.
- Support weight loss when consumed as portion-controlled snacks.

Possible Side Effects:
- Common cause of food allergies; avoidance is necessary for allergic individuals.
- Walnut hulls may cause skin irritation.

Quantity Recommendations:
- One ounce (about seven walnut halves) daily.
- Include walnuts in a variety of dishes such as smoothies, cereal, oatmeal, pasta, and salads.

PECAN

Benefits:
- Good source of manganese, which boosts the immune system and protects nerve cells.
- Effective in reducing the risk of heart problems by lowering bad cholesterol and increasing HDL.
- Aid in weight loss due to B-complex vitamins that boost metabolism.
- Prevent skin problems with vitamin A and zinc.

Possible Side Effects:
- Can trigger allergic reactions in sensitive individuals.

Quantity Recommendations:
- Consume a handful of pecans daily, about 20 kernels.
- American Heart Association recommends four servings of unsalted, unoiled nuts per week (1/3 cup).

BRAZILIAN NUTS

Benefits:

- Reduce the risk of heart disease due to antioxidants, selenium, vitamin E, magnesium, fiber, and Omega 3.
- Promote hair and nail health with selenium, zinc, and vitamins E and B.
- Support brain health and prevent cognitive decline due to anti-inflammatory and antioxidant properties.

Possible Side Effects:

- Excessive consumption can cause selenium toxicity, leading to stomach pain, nausea, and fatigue.
- Can cause gastrointestinal symptoms and allergic reactions.

Quantity Recommendations:

- 1 to 2 Brazil nuts daily, about five days a week.
- Consume without salt for maximum benefits.

CHESTNUTS

Benefits:

- Good source of fiber for digestive health.
- Contain antioxidants like vitamin C and phytonutrients that protect cells from damage.
- Provide copper, essential for blood vessel formation.

Possible Side Effects:

- Rare allergic reactions causing wheezing and throat swelling.

Quantity Recommendations:
- Consume small to moderate quantities daily.
- Ten chestnuts provide about 17% of the recommended daily value of fiber.

PEANUTS

Benefits:
- High protein content supports tissue repair and building.
- Maintain cardiovascular health with unsaturated fatty acids.
- Protect against heart disease with resveratrol, a phytonutrient with antioxidant properties.
- Help lower cholesterol with phytosterols.
- Aid in weight loss by controlling blood sugar levels and increasing satiety.
- May prevent type 2 diabetes and some cancers with phytosterols and nutrients.

Possible Side Effects:
- Common cause of allergic reactions.
- Prone to contamination with aflatoxins and nutrient inhibitors like phytate.

Quantity Recommendations:
- A serving size is about 30 grams of roasted peanuts.
- Consume up to 30 grams daily, ensuring diet diversity.

PEANUT BUTTER

Benefits:

- **Heart Disease Reduction:** Contains resveratrol, an antioxidant that helps fight free radicals causing heart disorders and cancer.
- **Weight Loss:** Suppresses hunger by combining fiber and protein, making you feel full for longer.
- **Gallstone Prevention:** Can lower bad cholesterol and increase good cholesterol, reducing the possibility of developing gallstones.
- **Bone Health:** Abundant in iron and calcium, promoting strong bones.

Possible Side Effects:

- **Allergic Reactions:** Should be avoided by individuals with nut allergies due to the risk of severe adverse reactions.
- **Saturated Fats:** Contains saturated fats, which can lead to heart disorders if consumed excessively.
- **Phosphorus:** Can prevent the proper absorption of other minerals, such as zinc and iron.

Quantity Recommendations:

- **General Consumption:** Two tablespoons daily provide approximately 180 calories. More than this can lead to undesired weight gain.
- **Muscle Growth & Immunity:** High protein content is essential for muscle growth and immune boosting.

- **Controlled Portions:** Due to its high-calorie content, one tablespoon providing about 100 kilocalories and 10 grams of fat is recommended.

PEANUT OIL

Benefits:

- **Blood Sugar Control:** High in unsaturated fats, boosting insulin sensitivity and maintaining blood sugar levels.
- **Hair & Skin Health:** Vitamin E content strengthens hair follicles, reduces damage, and prevents dandruff.
- **Anti-aging:** Helps reduce dark spots, wrinkles, pigmentation, and fine lines due to its high vitamin E content.
- **Joint Health:** Reduces inflammation in arthritis and strengthens joints.
- **Heart Health:** Contains healthy cholesterol, which helps reduce harmful cholesterol levels.

Possible Side Effects:

- **Allergic Reactions:** Can cause an allergic reaction in people with peanut allergies.
- **High-Heat Cooking:** Not suitable for high-heat cooking due to its low smoke point, making it more susceptible to oxidation.

Quantity Recommendations:

- **General Consumption:** About 1.5 tablespoons of peanut oil daily, focusing on replacing fats and oils higher in saturated fats.
- **Heart Health:** High in vitamin E and polyunsaturated fats, contributing to a lower risk of heart disease.

- **Moderation:** Due to the high content of omega-6 fatty acids, moderation is recommended.

SAFFLOWER OIL

Benefits:

- **Heart Health:** Prevents heart diseases like atherosclerosis and stroke.
- **Muscle Health:** Improves muscle strength, especially during menopause.
- **Anti-inflammatory:** Rich in monosaturated and polysaturated fat acids, enhancing anti-inflammatory effects.

Possible Side Effects:

- **Gastrointestinal Issues:** May cause vomiting, diarrhea, and stomachache.
- **Allergic Reactions:** Can trigger allergies in some people.

Quantity Recommendations:

- **General Consumption:** Recommended safe intake is around 1 3/4 tablespoons daily.
- **Blood Sugar Control:** Improves blood sugar levels due to high unsaturated fats.

SUNFLOWER SEEDS

Benefits:

- **Nutrient-Rich:** Source of proteins, fats, vitamins, and minerals, delaying age-related diseases like cardiovascular diseases and cognitive impairment.

- **Antioxidants:** Rich in carotenoids and polyphenols, protecting against various cancers.
- **Inflammation Reduction:** High in monounsaturated fat, reducing inflammation and CRP levels.
- **Digestive Health:** Fiber-rich, beneficial for digestion and weight management.

Possible Side Effects:

- **Allergic Reactions:** People allergic to peanuts may also be allergic to sunflower seeds.
- **Toxicity:** Excessive consumption may lead to kidney damage due to cadmium traces.

Quantity Recommendations:

- **General Consumption:** Around 30 grams, or a quarter cup, daily for cardiovascular protection and antioxidant benefits.
- **Moderation:** Include in small amounts as part of the daily fat intake.

PUMPKIN SEEDS

Benefits:

- **Heart Health:** Contains polyunsaturated fatty acids like omega-3 and omega-6, sterols, and phytosterols, maintaining heart health.
- **Bone Health:** Rich in magnesium, essential for bone formation.

Possible Side Effects:

- **Digestive Issues:** High fiber content may cause bloating or flatulence if consumed excessively.

Quantity Recommendations:

- **General Consumption:** About 120 grams (4.2 oz) daily for optimum health benefits.
- **Weight Management:** Include in the diet for increased satiety and muscle mass support.

PUMPKIN

Benefits:

- **Cancer Protection:** Contains Beta-carotene, reducing cancer risk and protecting against heart disorders and macular degeneration.
- **Blood Pressure Regulation:** High potassium content helps lower blood pressure.

Possible Side Effects:

- **Diuretic Effect:** May increase urine output if consumed in large amounts.

Quantity Recommendations:

- **General Consumption:** One serving daily to meet fiber, vitamin, and mineral intake recommendations.
- **Eye Health:** Source of vitamin A, essential for vision.

WATERMELON

Benefits:

- **Hydration:** Composed of 92% water, keeping the body well-hydrated.
- **Cardiovascular Health:** Reduces body weight, BMI, waist-hip ratio, and blood pressure.

- **Antioxidant Properties:** Rich in lycopene, vitamin A, and vitamin C, providing antioxidant potential.

Possible Side Effects:

- **Digestive Discomfort:** Excessive consumption can lead to bloating, constipation, and diarrhea due to high FODMAP content.

Quantity Recommendations:

- **General Consumption:** Four servings of fresh fruits daily, with watermelon being one of them.

PINEAPPLE

Benefits:

- **Anti-inflammatory:** Bromelain helps lower inflammation.
- **Immune System Boost:** High in vitamin C, defending the body from germs.
- **Bone Health:** Contains abundant manganese, essential for bone formation.
- **Tissue Healing:** Bromelain promotes skin and tissue healing.

Possible Side Effects:

- **Mouth Tenderness:** Excessive consumption can cause mouth tenderness due to its meat-tenderizing properties.
- **Vitamin C Overdose:** May lead to nausea, diarrhea, vomiting, abdominal pain, or heartburn.

Quantity Recommendations:

- **General Consumption:** Two slices (diced) daily, providing approximately 100 mg of ascorbic acid.
- **Balanced Diet:** Two cups of sliced pineapple daily, together with other fruits.

PAPAYA

Benefits:

- **Digestive Health:** High fiber content reduces toxin absorption, facilitating intestinal function.
- **Weight Loss:** Low glycemic index and high fiber content promote satiety.
- **Cholesterol Reduction:** Lowers LDL cholesterol and inhibits cholesterol oxidation in arteries.

Possible Side Effects:

- **Pregnancy Risk:** Green papaya is contraindicated for pregnant women due to its high latex content.
- **Diarrhea Risk:** Ripe papaya should be eaten in moderation to avoid diarrhea.

Quantity Recommendations:

- **General Consumption:** Medium slice daily, can be consumed every day for digestive health benefits.
- **Antioxidant Effect:** Regular consumption provides powerful antioxidant benefits.

AVOCADO

Benefits:

- **Nutrient-Rich:** Good source of vitamin K and potassium, essential for blood clotting and heart health.
- **Weight Management:** High fiber content contributes to weight loss and reduces blood sugar spikes.
- **Heart Health:** High in monounsaturated oleic fats, essential for lipid level control.

Possible Side Effects:

- **Weight Gain:** Excessive consumption can lead to weight gain due to high fat content.
- **Digestive Distress:** Overeating can cause digestive issues like diarrhea and bloating.

Quantity Recommendations:

- **General Consumption:** Half an avocado daily, providing about 120 calories.
- **Balanced Diet:** Include in daily diet for brain, heart, and gut health benefits.

CARROT

Benefits:

- **Vision Improvement:** High in vitamin A, essential for vision, growth, and reproduction.
- **Cancer Prevention:** Rich in beta-carotene, preventing various cancers.

- **Weight Loss:** Low calorie and high fiber content promote satiety and weight loss.

Possible Side Effects:
- **Skin Discoloration:** Excessive consumption can lead to yellow or orange skin coloration.

Quantity Recommendations:
- **General Consumption:** Five servings of fresh vegetables daily, approximately half a cup of carrots.

BROCCOLI

Benefits:
- **Cardiovascular Health:** Rich in soluble fiber, potassium, magnesium, sulforaphane, and antioxidants.
- **Digestive Health:** Contains sulforaphane and fibers that prevent ulcers and constipation.
- **Blood Sugar Control:** Fiber content helps lower blood sugar levels.

Possible Side Effects:
- **Digestive Issues:** May cause bloating and gas in individuals with inflammatory bowel disease.
- **Thyroid Function:** Excess consumption can block iodine absorption.

Quantity Recommendations:
- **General Consumption:** About 150 grams daily for cancer prevention.

BRUSSELS SPROUTS

Benefits:

- **Antioxidants:** High in antioxidants, nutrients, and fiber, protecting against cancer.
- **Weight Control:** Helps with weight management due to high fiber content.

Possible Side Effects:

- **Medication Interaction:** Can interact with blood thinner medications.
- **Digestive Issues:** Excessive consumption may cause obstipation and stomachache.

Quantity Recommendations:

- **General Consumption:** 3 to 4 servings of vegetables and dark greens daily.

KALE

Benefits:

- **Cancer Prevention:** Potent inhibitory activities against cancer cell lines due to phytochemicals.
- **Gut Health:** Increases beneficial bacterial diversity and reduces inflammation.

Possible Side Effects:

- **Kidney Dysfunction:** High potassium content can cause hyperkalemia.

- **Medication Interference:** Can interfere with beta-blockers and blood thinners.

Quantity Recommendations:

- **General Consumption:** One to two cups daily, depending on health condition and medication.

SPINACH

Benefits:

- **Eye Health:** High levels of chlorophyll and carotenoids, essential for maintaining healthy eyesight.
- **Blood Quality:** Improves blood quality and restores energy levels due to rich iron content.
- **Heart Health:** Rich in nitrates, improving blood flow and pressure.

Possible Side Effects:

- **Kidney Stones:** High oxalate content may contribute to kidney stones.
- **Medication Interaction:** High Vitamin K content may interact with blood-thinning medication.

Quantity Recommendations:

- **General Consumption:** 1/2 cup cooked or 1 cup raw daily.

LETTUCE

Benefits:

- **Mineral Source:** Good source of minerals, maintaining physiological functions.

- **Low Calories:** Helps with weight loss due to low-calorie content.
- **Sleep Aid:** Contains sleep-potentiating compounds, improving sleep quality.

Possible Side Effects:
- **Allergic Reactions:** May cause oral allergy or anaphylaxis in rare cases.

Quantity Recommendations:
- **General Consumption:** Daily consumption in salads, soups, or smoothies. A serving is 1 cup of raw, green leafy vegetables.

Each food listed provides unique health benefits and potential side effects. By following the recommended quantities, you can enjoy these foods as part of a balanced diet, contributing to overall health and well-being.

TOMATO

Overview: The tomato is the edible berry of Solanum lycopersicum, commonly known as a tomato plant. The species originated in western South America and Central America.

Benefits:
- **Cancer Prevention:** Contains lycopene, which has anticancer properties, particularly against prostate cancer.
- **Cardiovascular Health:** Rich in antioxidants and dietary fiber, helping to lower LDL cholesterol and maintain healthy blood vessels.

- **Immune System Boost:** Source of vitamin C, strengthening the body's natural defenses.
- **Vision, Skin, and Hair Health:** Rich in carotenoids that transform into vitamin A, essential for good vision, shiny hair, and well-hydrated skin.

Possible Side Effects:

- **Allergic Reactions:** Contains histamine, which may trigger skin rashes and allergic reactions.
- **Kidney Issues:** High potassium content should be avoided by patients with chronic renal dysfunction due to the risk of kidney stones.

Quantity Recommendations:

- **General Consumption:** One medium tomato or seven cherry tomatoes daily. The recommended intake is five servings per day, with one serving being approximately one medium tomato. Lycopene content is higher when consumed as juice or paste.

CABBAGE

Overview: Cabbage is a leafy vegetable from the Brassica oleracea family, available in green, red (purple), or white (pale green) varieties.

Benefits:

- **Blood Pressure Control:** Rich in fiber, folic acid, and potassium, which helps relax blood vessels and lower blood pressure.
- **Weight Loss:** Low in calories and high in dietary fiber, promoting satiety.

- **Vision Health:** Rich in vitamin A, reducing the risk of macular degeneration.
- **Immune System Support:** Source of vitamin C, carotenoids, and folate, which prevent free radical damage and infections.
- **Digestive Health:** High fiber content helps prevent constipation.
- **Anti-inflammatory:** Contains antioxidants that fight inflammation and prevent chronic diseases.

Possible Side Effects:

- **Gas and Flatulence:** High sulfur and raffinose content can lead to increased gas.
- **Medication Interaction:** Can interact with blood thinners due to its vitamin K content.

Quantity Recommendations:

- **General Consumption:** About one cup of cabbage daily, as part of the recommended 2-3 cups of cruciferous vegetables.

POTATO

Overview: The potato is a starchy tuber of Solanum tuberosum, native to the Americas.

Benefits:

- **Weight Loss:** High fiber content promotes fullness.
- **Anxiety Reduction:** Contains tryptophan, a natural sedative that helps calm nerves and promote sleep.

- **Heart Health:** Free of cholesterol and contains anthocyanins that fight oxidative stress.
- **Brain Health:** Contains alpha-lipoic acid, zinc, phosphorus, and B-complex vitamins, enhancing cognitive health.

Possible Side Effects:

- **High Glycemic Index:** Diabetics and those with cardiovascular diseases should consult healthcare professionals before increasing intake.

Quantity Recommendations:

- **General Consumption:** One medium-sized potato daily.

SWEET POTATO

Overview: Sweet potatoes are a starchy, sweet-tasting root vegetable from the Convolvulaceae family.

Benefits:

- **Vitamin A Deficiency Prevention:** High in beta-carotene, converted into vitamin A.
- **Diabetes Management:** Low glycemic index helps control blood sugar levels.
- **Stress Reduction:** High magnesium content helps reduce stress and anxiety.
- **Anti-inflammatory:** Contains vitamins with powerful anti-inflammatory properties.

Possible Side Effects:

- **Skin Discoloration:** Excessive consumption may cause slight orange discoloration of the skin and nails.

Quantity Recommendations:
- **General Consumption:** Include sweet potatoes in the diet weekly to meet nutrient intake recommendations.

MAIZE/CORN

Overview: Maize, also known as corn, is a cereal grain first domesticated in southern Mexico.

Benefits:
- **Eye Health:** High in lutein, zeaxanthin, and beta-carotene, protecting against oxidative damage.
- **Cardiovascular Health:** Corn oil contains essential fatty acids and vitamin E, preventing oxidative stress and atherosclerosis.
- **Weight Loss:** Contains resistant starch, a dietary fiber that reduces food intake and modulates gene expression.

Possible Side Effects:
- **Fungal Contamination:** Susceptible to mycotoxins.
- **Gluten Intolerance:** Contains zein and gluten proteins.

Quantity Recommendations:
- **General Consumption:** Less than 450 grams per day, with a serving being around 100 grams.

RICE

Benefits:
- **Weight Loss:** High fiber content promotes bowel movements and lowers cholesterol.

- **Cancer Prevention:** Contains antioxidants that prevent cellular damage.

Possible Side Effects:
- **Allergies:** Rare but possible, with symptoms similar to wheat and gluten allergies.

Quantity Recommendations:
- **General Consumption:** One cup per serving, primarily if it is the main carbohydrate source.

WHEAT

Benefits:
- **Source of Carbohydrates:** Promotes satiety and provides essential nutrients.
- **Colon Cancer Risk Reduction:** Whole grain wheat is high in fiber.

Possible Side Effects:
- **Celiac Disease:** Must be avoided by those with celiac disease or gluten intolerance.

Quantity Recommendations:
- **General Consumption:** 5 to 8 ounces of grains daily, with 3 to 6 ounces being whole grains.

SOYBEAN

Benefits:
- **Cardiovascular Health:** Rich in omega-3 and isoflavones, lowering cholesterol and triglycerides.

- **Menopause Symptom Relief:** Isoflavones help regulate hormone levels.
- **Bone and Skin Health:** Reduces calcium elimination and stimulates collagen production.

Possible Side Effects:
- **Allergies:** Recognized as a significant food allergen.

Quantity Recommendations:
- **General Consumption:** About 85g of cooked soy, 30g of tofu, or 1 cup of soy milk daily.

BEANS

Benefits:
- **Chronic Disease Reduction:** Contains phytonutrients with antioxidant properties.
- **Weight Loss:** High fiber content promotes fullness and controls cholesterol levels.
- **Diabetes Management:** Provides a slow source of glucose.

Possible Side Effects:
- **Migraines:** Contains tyramine, which can trigger migraines.
- **Flatulence:** Causes gas build-up when breaking down sugars.

Quantity Recommendations:
- **General Consumption:** At least one cup of cooked beans daily.

ZUCCHINI

Benefits:

- **Digestive Health:** High fiber content promotes healthy digestion.
- **Nutrient-Rich:** Rich in vitamins, minerals, and antioxidants.
- **Vision Health:** Contains lutein and zeaxanthin.

Possible Side Effects:

- **Kidney Stones:** High oxalate content may cause kidney stones.
- **Allergies:** Pesticide content can cause allergic symptoms.

Quantity Recommendations:

- **General Consumption:** About 2 to 3 cups daily for digestive benefits and nutrient intake.

These recommendations provide a balanced approach to incorporating these foods into a healthy diet, considering their benefits, potential side effects, and appropriate quantities for consumption.

OTHERS

Chicken

Chicken helps maintain a healthy weight as it is a protein source promoting satiety. This food is also a source of B-complex vitamins, particularly vitamin B3 (niacin), which is essential for brain and energy metabolism. Chicken is beneficial for physical performance, recovery, strength, and endurance due to its vitamins, minerals, and bioactive compounds like anserine, carnosine, and creatine.

Possible side effects include allergic reactions, which are rare but can affect both adults and children. Thoroughly cooked chicken is essential to avoid salmonella, a bacterium that can cause diarrhea, fever, pain, and stomach cramps.

Chicken meat can be consumed daily for lunch and dinner or as a snack, but fried chicken should be avoided. The healthiest cut is the breast. A medium fillet at lunch or dinner is typically sufficient to meet nutritional needs. It is recommended to consume one to two servings (80 grams of cooked lean chicken) per week, with non-lean cuts like thighs or wings consumed less frequently.

Barley

Barley is a good fiber source, promoting digestive health and weight loss by enhancing satiety and better glycemic control. It is rich in minerals such as manganese, selenium, and phosphorus, essential for thyroid, bone, and teeth health. Barley's phytochemicals with antioxidant and anti-inflammatory properties reduce the risk of heart disease and cancer.

Individuals with irritable bowel syndrome should consult healthcare practitioners before increasing barley intake due to its fructans content, which can cause bloating and gas. Those with celiac disease or wheat allergies should avoid barley as it contains gluten.

Consume about 100 grams of barley daily, equivalent to one cup. Barley can reduce cardiovascular disease risk and maintain a healthy weight due to its dietary fiber content. It also supports brain health with its thiamine and selenium content.

Bananas

Bananas are rich in vitamins B6 and C, helping metabolize macronutrients and boosting the immune system. They regulate blood pressure due to their potassium content, providing about 10% of the recommended daily value.

Overconsumption may lead to high blood potassium levels, a concern for people with blood pressure issues. Individuals with severe renal disease should monitor banana consumption due to its potassium content.

A medium-sized banana is the serving size, and it is recommended to consume two to four servings of fruit daily. Bananas can be eaten daily, particularly for breakfast or as an afternoon snack, but those with diabetes should control portions.

Mango

Mango is a rich source of vitamin C, boosting the immune system. One cup provides nearly 70% of the recommended daily vitamin C intake. Mangoes contain polyphenols with antioxidant properties that reduce the risk of chronic diseases.

Excessive consumption may cause gastrointestinal distress, so individuals with irritable bowel movements should be cautious. Mango contains Urushiol, an oily substance that can cause allergic reactions in some people. Diabetics should consume mangoes in moderation due to their fructose content.

It is recommended to consume one to two servings of mango daily, equating to one and a half to two cups of diced mango. Mango is high in calories, so moderate consumption is advised to avoid weight gain.

Orange

Oranges reduce stroke risk due to their flavonoid and vitamin C content, which have powerful antioxidant effects. They strengthen the immune system with vitamins A, B, C, and folate and promote intestinal health with fibers like pectin, cellulose, and hemicellulose.

Individuals with heartburn may experience worsening symptoms due to oranges' organic acids. Oranges should be consumed with their fiber for maximum benefits. At least one raw orange or 150 mL of natural juice should be consumed daily.

Oranges help with weight loss and provide high vitamin C content. They should be included in the daily fruit intake, with an average recommendation of 3-5 portions of fruit daily. Limit orange juice to no more than 8 ounces (240 ml) per day.

Apple

Apples support gut health as a fiber source, reduce sweet cravings, and provide polyphenols that protect against cell damage. Some people may be allergic to apples, and it's best to avoid them if experienced.

Apples can be eaten daily to improve satiety and gut function and reduce blood sugar. One apple a day is recommended as part of a healthy diet. Apples are rich in soluble fiber, promoting healthy gut bacteria and potentially preventing heart disease and cancer.

Strawberry

Strawberries are packed with phytonutrients and antioxidants that protect against cancer, cardiovascular diseases, and inflammation. They have a low glycemic index and regulate blood sugar levels. Strawberries are also a good source of vitamin C, potassium, and folic acid.

Possible side effects include pesticide residues and adverse effects for individuals on Beta-blockers. Individuals with pollen-food allergies or sensitivities to salicylates should be cautious.

It is recommended to consume eight strawberries a day. Diabetic patients should include strawberries in their diet to help reduce complications. Strawberries can be consumed in various ways, ideally two to three units a day.

Blueberry

Blueberries are high in antioxidants, particularly anthocyanins, which manage cardiovascular diseases, inflammatory responses, different types of cancer, and microbial activity. They are rich in vitamins A, D, E, and folic acids.

Possible side effects include allergic reactions and pesticide residues. Those with a Glucose-6-phosphate dehydrogenase (G6PD) deficiency should avoid blueberries.

A serving of fresh blueberries is approximately 150 grams, and they can be consumed daily. One cup of blueberries contains 84kcal, 1.1g of protein, 0.49g of fat, 21.45g of carbohydrates, and 3.6g of fiber.

Cranberries

Cranberries manage urinary tract infections with their high proanthocyanin content. They reduce cardiovascular disorder risks with their polyphenols and have antioxidant properties that fight cell damage.

Excessive consumption can cause stomach irritation, diarrhea, and increase kidney stone risk. Cranberries may interfere with blood-thinning medications due to their vitamin K content.

It is recommended to consume 15 grams of cranberries daily. Cranberries can be used in various preparations or consumed as juice. Whole, fresh cranberries are also a healthy choice.

Cherries

Cherries reduce gout risks by lowering uric acid levels and inflammation. They contain antioxidants that reduce oxidative stress and inflammation and lessen muscle soreness.

Individuals with irritable bowel syndrome should consult healthcare professionals before increasing cherry intake. Cherries can cause digestive discomfort due to their sugar content.

It is recommended to consume 25 cherries a day. Cherries can be consumed raw or used in various dishes. Daily consumption of cherry juice can reduce inflammation and chronic disease risks.

Pear

Pears are rich in antioxidants, reducing cancer risk and inflammatory diseases. They help reduce obesity due to their low calorie and high water and fiber content and are a good copper source for immune response.

Possible side effects include allergic reactions, though pears have a lower allergy risk. Excessive fiber intake may cause abdominal discomfort.

Pears can be eaten once a day and are low in calories, supporting weight loss. One medium pear is a serving, with a recommendation of at least two servings or two cups of fruit per day.

Grape

Grapes reduce inflammation and prevent chronic diseases with their antioxidants, including resveratrol, vitamin C, and beta-carotene. They support cardiovascular health and lower hypertension risk.

Overeating grapes may cause gastrointestinal symptoms, such as bloating and diarrhea. Those with diabetes and kidney problems should consume grapes in moderation.

A serving of grapes is a cup, and it is recommended to consume two to four servings of fruit daily. Grapes are beneficial for heart health, eye health, and managing type 2 diabetes.

Chickpea

Chickpeas lower cardiovascular disease risk with their high fiber content, low glycemic index, and phytosterols. They help control diabetes, prevent certain cancers, aid in weight loss, and improve bowel health.

Possible side effects include intestinal gas due to oligosaccharides and potassium content, requiring caution for patients with kidney dysfunctions.

It is recommended to consume 150 grams of chickpeas, with a serving size of 1/4 cup of cooked chickpeas. Soaking chickpeas for about 12 hours before cooking can reduce bloating.

Pigeon Pea

Pigeon peas are a good protein source for vegans and vegetarians and are effective against anemia due to their iron content. They help control blood sugar with their fiber and antioxidants.

Possible side effects include allergic reactions and anaphylaxis. Pregnant women should avoid pigeon peas due to a lack of studies.

Pigeon peas can be consumed daily as part of a healthy meal.

Lentil

Lentils help maintain blood pressure and cholesterol levels due to their fiber, folate, and potassium content. They are excellent sources of plant proteins, helping build and repair body tissues.

Possible side effects include digestive symptoms such as bloating and flatulence due to high fiber content.

A serving of uncooked lentils is about one-quarter of a cup. Eating a serving of lentils daily is recommended.

Split Pea

Split peas improve digestion with their high dietary fiber content, prevent iron deficiency anemia, and are excellent sources of vegetable protein. They keep eyes healthy with their lutein content.

Possible side effects include digestive issues such as gas, bloating, and diarrhea due to FODMAPs.

It is recommended to consume five servings per day of vegetables, including split peas. A serving size of cooked split peas is 1/4 cup.

Mung Bean

Mung beans contain vitamin K, improving blood clotting and bone health. They also contain isoflavones, which can prevent breast cancer and alleviate menopause symptoms.

Possible side effects include digestive issues for people with deficiency-cold in the spleen and stomach.

Mung beans can be consumed daily, with a recommendation of up to one cup daily. Varying vegetable intake is advisable.

Lima Beans

Lima beans lower glycemic index and control blood sugar with their dietary fiber and protein. They help reduce bad cholesterol and provide various vitamins and minerals.

Possible side effects include favism for individuals lacking the G6PD enzyme and digestive problems due to oligosaccharides.

A cup of raw ripe lima beans provides enough fiber to meet daily nutritional requirements. Lima beans contain ample amounts of dietary fiber and protein.

These summaries offer a concise overview of the benefits, possible side effects, and quantity recommendations for each food, helping you make informed dietary choices.Top of Form

Quinoa

Quinoa is a flowering plant in the amaranth family, grown primarily for its edible seeds. It is rich in protein, dietary fiber, B vitamins, and minerals, offering more significant amounts than many grains. It is an excellent plant-based protein source for vegans due to its high amino acid content. Its high fiber and low glycemic index help control appetite and aid in weight loss. Quinoa's flavonoids (antioxidants) can prevent cellular damage, protecting the heart and brain.

Possible side effects of excess quinoa and low water intake include intestinal symptoms such as obstipation, stomach aches, itchy skin, and allergies. The daily fiber recommendation is about 25-30g, so around half a cup of quinoa daily is recommended, along with a variety of other fibers and increased water intake.

Chia Seed

Chia, or Salvia hispanica, is a species of flowering plant in the mint family, native to central and southern Mexico and Guatemala. It is a pseudocereal, cultivated for its edible, hydrophilic seeds. Chia seeds help detoxify the body due to their fiber and antioxidant content and are rich in omega-3 fatty acids, which may help reduce heart disease risk.

Possible side effects include allergic reactions, with symptoms such as dermatitis or anaphylaxis. It is recommended to consume about one and a half tablespoonfuls of chia seeds twice a day for optimum health benefits. Consult a registered dietitian for individualized advice if you have underlying health conditions.

Brown Rice

Brown rice is a whole grain with the inedible outer hull removed, retaining the bran layer and the cereal germ. It reduces the risk of heart disease due to its high fiber and magnesium content, which can help prevent heart disease and stroke. Brown rice aids in weight loss as higher-fiber foods cause longer satiety with fewer calories.

Possible side effects include reduced nutrient absorption due to the antinutrient phytic acid, making digestion more difficult. Combining rice with vegetables, vinegar, or oil can lower its glycemic index (GI), which measures how quickly a food increases blood sugar.

Jasmine Rice

Jasmine rice is a long-grain aromatic rice variety with a fragrance reminiscent of pandan and popcorn. Brown jasmine rice is rich in insoluble fibers, helping prevent constipation and normalize intestinal rhythm. It is a source of B-complex vitamins, especially folic acid, which is crucial for a healthy pregnancy and fetal development. Brown jasmine rice also helps reduce blood glucose levels, benefiting individuals with type 2 diabetes due to its high fiber content.

White jasmine rice should be consumed in moderation by diabetic patients due to its low fiber content and high glycemic index. Jasmine rice typically contains lower arsenic levels than most rice grown worldwide. The recommendation for adult grains such as jasmine rice is six servings daily, with a serving size of approximately half a cup of cooked rice. Whole grains are healthier than refined grain products.

Wild Rice

Wild rice comprises four species of grasses in the genus Zizania, historically gathered and eaten in North America and China. It is a good source of carbohydrates, providing energy and sparing protein for other functions. Wild rice is rich in fiber, helping control calorie intake and maintain blood glucose and cholesterol. It contains the antioxidant apigenin, which may help prevent cancer and oxidative damage.

Wild rice may contain heavy metals such as arsenic and other toxins, so it must be adequately cooked to minimize this risk. The serving size of wild rice is one-quarter of a raw portion. About six to eleven servings daily in the cereal group is recommended, incorporating other cereals and carbohydrate sources for dietary diversity.

Arborio Rice

Arborio rice is an Italian short-grain rice named after the town of Arborio in Italy's Po Valley. The rounded grains are firm, creamy, and chewy when cooked due to their higher amylopectin starch content. Arborio rice is an excellent carbohydrate source, providing energy and sparing protein for other biochemical processes. It is also rich in dietary fiber, helping maintain a healthy weight and colon health.

Like most grains, arborio rice is rich in carbohydrates, so overeating can lead to unhealthy weight gain. About one-third of an uncooked portion of arborio rice is a serving. As a grain, six to eleven servings daily are recommended according to the food guide pyramid. Dietary diversity is essential for healthy eating.

BLOOD GROUP "A" FOODS 1

MEATS & POULTRY	SEAFOOD	EGGS & DAIRY
Highly Beneficial None.	**Highly Beneficial** Carp, Cod, Grouper, Mackerel, Monkfish, Pickerel, Red Snapper, Rainbow Trout, Salmon, Sardine, Sea Trout, Silver Perch, Snail, Whitefish, Yellow Perch.	**Highly Beneficial** Soya Cheese Soy Milk (Good Dairy alternatives).
Neutral Chicken, Cornish, Hens, Turkey.	**Neutral** Abalone, Albacore (Tuna), Mahi-mahi, Ocean Perch, Pike, Porgy, Sailfish, Sea Bass, Shark, Smelt, Snapper, Sturgeon, Swordfish, Weakfish, White Perch, Yellowtail.	**Neutral** Farmer, Feta, Goat cheese, Goat milk, Kefir Mozzarella: low fat, Ricotta: low fat, String cheese, Yogurt: regular, W/fruit, frozen, eggs.

MEATS & POULTRY	SEAFOOD	EGGS & DAIRY
Avoid	Avoid	Avoid
Bacon, Beef, regular, ground Buffalo, Duck, goose, Ham, Heart, Lamb, Liver, Mutton, Partridge, Pheasant, Pork, Quail, Rabbit, Veal Venison	Anchovy, Barracuda, Betuga, Bluefish, Bluegill Bass, Catfish, Caviar, Clam, Conch, Crab, Crab, Eel, Flounder, Frog, Grey Sole, Haddock, Hake, Halibut, Herring: fresh/, Lobster, Lox (smoked salmon), Mussels, Octopus, Oysters, Scallop, Shad, Shrimp, sole, Squid (calamari), Striped Bass, Tilefish, Tur.tle	American cheese, Blue cheese, Brie, Butter, Buttermilk, Camembert, Casein, Cheddar, Colby, Cottage, Cream cheese, Edam, Emmenthal, Gouds, Gruyere, Ice cream, Jarisberg, Monterey jack, Munster, Parmesan, Provolone, Neufchatel, Sherbet Skim or 2% milk, Swiss, Whey, Whole milk

BLOOD GROUP "A" FOODS 2

OILS & FATS	NUTS & SEEDS	BEANS & LEGUMES	CEREALS
Highly Beneficial Linseed (flaxseed) oil, Olive oil	**Highly Beneficial** Peanuts, Peanut butter, Pumpkin seeds	**Highly Beneficial** Beans: adzuke, azuki, black, green, pinto, red soy Lentils: domestic, green, red Peas: black-eyed.	**Highly Beneficial** Amaranth, Buckwheat, Kasha

OILS & FATS	NUTS & SEEDS	BEANS & LEGUMES	CEREALS
Neutral Canola oil, Cod liver oil.	**Neutral** Almond butter Nuts: almonds, chestnuts, filberts, hickory, litchi, macadamia, pignola (pine), Poppy seeds, Sesame seeds Sesame butter (tahini), Sunflower butter, Sunflower, seeds, Walnuts.	**Neutral** Beans, cannellini, broad, fava, jicama, snap, string, white Peas: green, pods, snow.	**Neutral** Barley, Cornflakes, Cornmeal, Cream of rice, Kamut, Millet: puffed Oat bran, Oatmeal, Rice: puffed, Rice bran, Spelt
Avoid Corn oil, Cottonseed oil, Peanut oil, Safflower oil, Sesame oil.	**Avoid** Brazil nuts, Cashews, Pistachios.	**Avoid** Beans: copper, garbanzo, kidney, lima, navy, red, tamarind.	**Avoid** Cream of wheat, Familia, Farine Granola, Grape nuts, wheat germ, Seven grain, Shredded wheat, Wheat bran

BLOOD GROUP "A" FOODS 3

BREADS & MUFFINS	GRAINS & PASTAS	VEGETABLES
Highly Beneficial Essene bread, Ezekiel bread, Rice cakes, Soya flour bread, Spouted wheat bread.	**Highly Beneficial** Buckwheat Kasha Flour: oat, rice rye, soba, Pasta, artichoke	**Highly Beneficial** Artichoke: domestic, Jerusalem, Beet leaves, Broccoli, Carrots, Chicory, Collard greens, Dandelion, Escarole, Garlic, Horseradish, Kale, Kohlrabi, Leek, Lettuce: romaine, Okra, Onions: red, Spanish, yellow, Parsley, parsnips, Pumpkin, spinach, Sprouts: alfalfa, Swiss chard, Tempeh, Tofu, Turnips.

BREADS & MUFFINS	GRAINS & PASTAS	VEGETABLES
Neutral	Neutral	Neutral
Bagels: wheat, Brown rice bread, Corn muffins, Fin Crisp, Gluten-free bread, Ideal Flat Bread, Millet, Oat bran muffins, Rye bread: 10% Rye Crisps, Rye Vita, Spelt bread, Wasa bread.	Couscous, Flour: barley, bulgur wheat, Durum wheat, gluten, graham, spelt, sprouted wheat. Noodles: spelt Quinoa Rice: basmati, brown, white, wild	Arugula, Asparagus, Avocado, Bamboo shoots, Beets, Bok Choy, Caraway, Cauliflower, celery, chervil, Coriander, Corn: white, yellow, cucumber, Dalkon radish, Endive, Fennel, Fiddlehead ferns, Lettuce: bibb, Boston, iceberg, mesclun, Mushroom: abalone, enoki, Portobello, tree oyster, Mustard greens, Olives: green, Onions: green Radicchio, Radishes, Rappini, Rappini, Scallion, Seaweed, Shallots, Sprouts: Brussels, mung, radish, Squash: all types, Water chestnut, Water cress, Zucchini

BREADS & MUFFINS	GRAINS & PASTAS	VEGETABLES
Avoid	Avoid	Avoid
Durum wheat, English muffins, High-protein bread, Matzos: wheat, Multi-grain bread, Pumpernickel, Wheat bran muffins, Whole Wheat bread.	Flour: white, whole wheat Pasta: semolina spinach.	Cabbage: Chinese, red, white, Eggplant, Lima beans, Mushroom: domestic, shiitake, Olives: black, Greek, Spanish, Peppers: green, red, jalapeno, yellow, Potatoes: sweet, red, white, Tomatoes, Yams.

Blood Group "A" Foods 4

FRUIT	JUICES & FLUIDS	CONDIMENTS
Highly Beneficial	Highly Beneficial	Highly Beneficial
Apricots, Blackberries, Blueberries, Cherries, Cranberries, Figs: dried, fresh. Grapefruit, Lemons, Pineapple, Plums: dark green, red. Prunes, Raisins.	Apricot Carrot Celery Cherry: black Grapefruit Pineapple Prune Water (with lemon)	Mustard.

FRUIT	JUICES & FLUIDS	CONDIMENTS
Neutral	**Neutral**	Neutral
Apples, Currants: black, red Dates, Elderberries, Gooseberries, Grapes: red Concord, green, black. Guava, Kiwi, Kumquat, Limes, Loganberries, Melons: canang, casaba, Christmas, Crenshaw, musk, Spanish, watermelon, Nectarines, Peaches, Pears, Persimmons, Pomegranates, Prickly pears, Raspberries, Star Fruit, carambola, Strawberries.	Apple, Apple cider, Cabbage, Cucumber, Cranberry, Grapefruit, Vegetable juice (corresponding to highlighted vegetables).	Jam (from acceptable fruits), Jelly (from acceptable fruits), Pickles: dill, sour, kosher, sweet Relish Salad dressing (low-fat, from acceptable fruits ingredients)

FRUIT	JUICES & FLUIDS	CONDIMENTS
Avoid	Avoid	Avoid
Bananas, Coconuts, Mangoes, Melons: cantaloupe, honeydew, Oranges, Papayas, Plantains, Rhubarb, Tangerines.	Orange, Papaya, Tomato	Ketchup, Mayonnaise, Worcestershire sauce.

Blood Group "A" Foods 5

SPICES	HERBAL TEAS	MISC BEVERAGES
Highly Beneficial	**Highly Beneficial**	**Highly Beneficial**
Barley malt, Blackstrap molasses, Garlic, Ginger, Miso, Soy sauce, Tamari.	Alfalfa, Aloe, Burdock, Chamomile, Echinacea, Fenugreek, Ginger, Ginseng, Green tea, Hawthorn, Milk thistle, Rose hips, Saint-John's-wort, Slippery elm, Valerain.	Coffee: regular, decaf Tea: green Wine: red

SPICES	HERBAL TEAS	MISC BEVERAGES
Neutral	**Neutral**	**Neutral**
Agar, Allspice, Almond extract, Anise, Arrowroot, Basil, Bay leaf, bergamot, Brown rice syrup, Cardamom, Carob, Chervil, Chives, Chocolate, Cinnamon, Clove, Coriander, Cornstarch, Corn syrup, Cream of tartar, Cumin, Curry, Dill, dulse, Honey, Horseradish, Kelp, Maple syrup, Marjoram, Mint, Mustard (dry), Nutmeg, Oregano, Paprika, Parsley, Peppermint, Pimiento, rice syrup, Rosemary, Saffron, Sage, Salt, Savory, Spearmint, Sugar: brown, white, Tamarind, Tapioca, Tarragon, Thyme, Turmeric, Vanilla.	Chickweed, Coltsfoot, Dandelion, Dong quai, Elder, Gentian, Goldenseal, Hops, Horehound, Licorice root, Linden, Mulberry, Mullein, Parsley, Peppermint, Raspberry leaf, Sage, Sarsaparilla, Senna, Shepherd's purse, Skullcap, Spearmint, Strawberry leaf, Thyme, Vervain, White birch, With oak bark, Yarrow.	Wine: white

SPICES	HERBAL TEAS	MISC BEVERAGES
Avoid	**Avoid**	**Avoid**
Capers, Gelatin: plain, Pepper: black ground, cayenne, peppercorn, red flakes, white, Vinegar: apple cider, white, balsamic, red wine, Wintergreen.	Catnip Cayenne, Corn silk, Red clover, Rhubarb.	Beer Liquor: distilled Seltzer water Soda: club, cola, diet, other Tea, black: regular. decaf

Blood Group "AB" Foods 1

MEATS & POULTRY	SEAFOOD	EGGS & DAIRY
Highly Beneficial	**Highly Beneficial**	**Highly Beneficial**
Lamb, Mutton, Rabbit, Turkey.	Albacore (Tuna), Cod, Grouper, Hake, Mackerel, Mahi-mahi, Monkfish, Ocean perch, Pickerel, Pike, Porgy, Rainbow, Trout, Red snapper, Sailfish, Salmon, Sardine, Sea trout, Shad, Snail, Sturgeon.	Cottage cheese, Farmer, Feta, Goat cheese, Goat milk, Kefir, Mozzarella, Ricotta, Sour cream (non fat), Yogurt.

MEATS & POULTRY	SEAFOOD	EGGS & DAIRY
Neutral	**Neutral**	**Neutral**
Live, Pheasant.	Abalone, Bluefish, Carp, Catfish, Caviar, Herring (fresh), Mussels, Scallop, Shark, Silver perch, Smelt, Snapper, Sole, Squid (calamari), Swordfish, Tilefish, Weakfish, Whitefish, White perch, Yellow perch.	Casein, Cheddar, Colby, Cream cheese, Edam, Emmenthal, Gouda, Gruyere, Jarisberg, Monterey jack, Munster, Neufchatel, Skim or 2% milk, Soy cheese, Soy milk, String cheese, Swiss, Whey.
Avoid Bacon	**Avoid**	**Avoid**
Beef ground, regular, Buffalo, Chicken Cornish Hens, Duck, Goose, Ham, Heart, Partridge, Pork, Quail, Veal, Venison.	Anchovy, Barracuda, Beluga, Bluegill, Clam, Conch, Crab, Crayfish, Eel, Flounder, Frog, Gray sole, Haddock, Halibut, Herring (pickled), Lobster, Lox (smoked salmon,), Octopus, Oysters, Sea bass, Shrimp, Striped bass, Turtle, Yellowtail.	American cheese, blue cheese, Brie, Butter, Buttermilk, Camembert, Ice cream, Parmesan, Provolone, Sherbet, Whole milk.

BLOOD GROUP "AB" FOODS 2

OILS AND FATS	NUTS AND SEEDS	BEANS & LEGUMES	CEREALS
Highly Beneficial Olive Oil	**Highly Beneficial** Chestnuts, Peanuts, Peanut butter, Walnuts	**Highly Beneficial** Beans: navy, pinto red, redsoy, Lintels: green	**Highly Beneficial** Millet, Oat bran, Oatmeal, Rice bran, Rice: puffed, Spelt
Neutral Canola oil, Cod liver oil, Linseed (flaxseed oil), Peanut oil.	**Neutral** Almond butter Nuts: almonds Brazil, cashews Hickory, Litchi, Macadamia, pignola Pistachio	**Neutral** Beans: Broad, Cannellini, Copper, Green, Jicama, Northern, Snap, String, Tamarind, White Lentils: domestic, Red Peas: green, Pods.	**Neutral** Amaranth, Barley, Cream of rice, Cream of wheat, Familla, Farina, Granola, Grape nuts, Seven-grain, Shredded wheat, Soy flakes, soy granules, Wheat bran, Wheat germ.

OILS AND FATS	NUTS AND SEEDS	BEANS & LEGUMES	CEREALS
Avoid	**Avoid**	**Avoid**	**Avoid**
Corn oil, Cottonseed oil, Safflower oil, Sesame oil, Sunflower oil.	Filberts, Poppy seeds, Pumpkin seed, Sesame butter (tahini), Sesame seeds, sunflower butter, Sunflower seeds.	Beans: Aduke, Azuki, Black, Fava, Kidney, Garbanzo, Lima Peas: Black-eyed.	buckwheat, cornflakes, cornmeal, Kamut, Kasha

BLOOD GROUP "AB" FOODS 3

BREADS & MUFFINS	GRAINS & PASTAS	VEGETABLES
Highly Beneficial Brown rice bread, Essene bread, Ezekiel bread, Fin crisp, Millet, rice cakes, 100% rye bread, Rye Crisps, Rye Vita, soy flour bread, sprouted wheat bread, Wasa bread.	**Highly Beneficial** Flour: oat, rice, rye, sprouted wheat, Rice: basmati, brown, white, wild.	**Highly Beneficial** Beet Leaves, Beets, Broccoli, Cauliflower, Celery, Collard greens, cucumber, Dandelion, Eggplant, Garlic, Kale, Mustard greens, Parsley, Parsnips, Potatoes: sweet, Sprouts: alfalfa, Tempeh, Tofu, Yams: all types.

BREADS & MUFFINS	GRAINS & PASTAS	VEGETABLES
Neutral	Neutral	Neutral
Bagels: wheat, Durum wheat, Gluten-free bread, High-protein bread, Ideal Flat Bread Matzos: wheat, Multi-Grain Bread, Oat bran muffins, Pumpernickel, Spelt bread, What bran muffins, Whole wheat bread.	Couscous, Flour: barley, bulgur wheat, Durum wheat, gluten, Graham, spelt, white, whole wheat, Pasta: semolina, spinach, Quinoa.	Arugula, Asparagus, Bamboo shoots, Bokchoy, Cabbage: Chinese, red, white, Caraway, Carrots, Chervil, chicory, Coriander, Daikon, Endive, Escarole, Fennel, fiddlehead ferns, ginger, Horseradish, Kohlrabi, Leek, Lettuce: Bibb, Boston iceberg, mesclun, romaine, mushroom: domestic, enoki, portobello, oyster, tree Okra, Olives: green, Greek, Spanish. Onions: green, red, Spanish, yellow, Potatoes: red, white. Pumpkin, Radicchio, Rappini, Rutabaga, Scallion, Seaweed, Shallots, Snow peas, Spinach, Sprouts: Brussels, Squash: all types, Swiss chard, tomato, Turnips, Water chestnut, Water cress, Zucchini

BREADS & MUFFINS	GRAINS & PASTAS	VEGETABLES
Avoid

corn muffins. | Avoid

Buckwheat kasha, Pasta; artichoke, Soba noodles. | Avoid

Artichoke, domestic, Jerusalem, Avocado, corn: white, yellow, Lima beans, Mushroom: abalone, shitake, Olives: black, Peppers: green, red jalapeno, yellow, Radishes, Sprouts: mung, radish. |

BLOOD GROUP "AB" FOODS 4

FRUIT	JUICES & FLUIDS	CONDIMENTS
Highly Beneficial

Cherries, Cranberries, Figs: dried, fresh. gooseberries, Grapes: black, Concord, green, red. Grapefruit, Kiwi, Lemons, Loganberries, Loganberries, Pineapples, Plums: dark, green, red. | Highly Beneficial

Cabbage, Carrot, Celery, Cherry: black, Cranberry, Grape, Papaya. | Highly Beneficial

NONE |

FRUIT	JUICES & FLUIDS	CONDIMENTS
Neutral Apples, Apricots, Blackberries, Blueberries, Boysenberries, currants: black, red. Dates, Elderberries, Kumquat, Limes, Melon, cantaloupe, canang, crenshaw, Christmas, casaba, honeydew, musk, Spanish, watermelon, Nectarines, Papayas, Peaches, Pears, Plantains, Prunes, Raisins, Raspberries, Strawberries, Tangerines.	**Neutral** Apple, apple cider, apricot, Cucumber, Grapefruit, Pineapple, Prune, Water (with lemon), Vegetable juice (corresponding with highlighted vegetables).	**Neutral** Jam (from Acceptable fruits), Jelly (from acceptable fruits), Mayonnaise, Mustard, Salad dressing (low-fat, from acceptable ingredients).
Avoid Bananas, Coconuts, Guava, Mangoes, Oranges, Persimmons, Pomegranates, Prickly pear, Rhubarb, Star fruit (carambola).	**Avoid** Orange	**Avoid** Ketchup, Pickles: dill, kosher sweet, sour Relish, Worcestershire, sauce.

BLOOD GROUP "AB" FOODS 4

SPICES	HERBAL TEAS	MISC BEVERAGES
Highly Beneficial	**Highly Beneficial**	**Highly Beneficial**
Curry, Garlic, Horseradish, Miso, Parsley.	Alfalfa, Burdock, Chamomile, Erchinaces, Ginger, Ginseng, Green tea, Hawthorn, Licorice root, Rose hips, Strawberry leaf.	Coffee: regular, decaf Tea: green.

SPICES	HERBAL TEAS	MISC BEVERAGES
Neutral	**Neutral**	**Neutral**
Agar, Arrowroot, Basil, Bay leaf, Bergamot, Bown rice syrup, Cardamon, Carob, Chervil, chive, Chocolate, cinnamon, Clove, coriander, Cream of tartar, cumin, Dill, Dulse, honey, Kelp, Maple syrup, Marjoram, Mint, Molasses, Mustard (dry), Nutmeg, Paprika, Peppermint, Pimiento, Rice syrup, Rosemary, Saffron, Sage, Salt, savory, Soysauce, Spearmint, Sugar: brown, white, Tamari, Tamarind, Tarragon, Thyme, Turmeric, Vanilla, Wintergreen.	Catnip, Cayenne, chickweed, Dandelion, Dong quai, Elder, Goldenseal, Horehound, Mulberry, Parsley, Peppermint, Raspberry leaf, Sage, Saint-John's wort, Sarsaparilla, Slippery elm, Spearmint, Thyme, Valerian Vervain, White birh, White oak bark, Yarrow, Yellow dock.	Beer, Seltzer water Soda: club, Wine: red, white.

SPICES	HERBAL TEAS	MISC BEVERAGES
Avoid	**Avoid**	**Avoid**
allspice, almond extract, Anise, Barley malt, Capers, conrstarch, Corn syrup, Gelatin: plain, Pepper: black, ground, cayenne, peppercorn, red glakes, white, Tapioca, vinegar: apple cider, Balsamic, White, Red wine.	Alow, Coltsfoot, corn silk, Fenugreek, Gentian, Hops, Linden, Mullein, Red clover, Rhubarb, Senna, Sherpherd's purse, Skullcap.	Liquor, distilled Soda: cola, diet, other Tea: black regular or decaf.

BLOOD GROUP "B" FOODS 1

MEATS & POULTRY	SEAFOOD	EGGS & DAIR
Highly Beneficial Lamb, Mutton, rabbit, Venison.	**Highly Beneficial** Cod, Flounder, Grouper, Haddock, Hake, Halibut, Mackerel, Mahi-mahi, Monkfish, Ocean perch, Pickerel, Pike, Porgy, Salmon, Sardine, Sea trout, Shad, Sole, Sturgeon, Sturgeon eggs, (caviar).	**Highly Beneficial** Cottage cheese, Farmer, Feta, Goat cheese, Goat milk, Kefir, Mozzarella, Ricotta, Skim or 2% milk, Yogurt: regular, w/fruit, frozen.
Neutral Beef: regular, ground, Buffalo, Liver, Pheasant, Turkey, Veal.	**Neutral** Abalone, Albacore 9Tuna, Bluefish, Carp, Catfish, Herring: fresh, pickled, Rainbow trout, Red snapper, Sailfish, Scallop, Shark, Silver perch, Smelt, Snapper, Squid (calamari), Swordfish, Tilefish, Weakfish, White perch, Whitefish, Yellow Perch.	**Neutral** Brie, Butter, Buttermilk, Camembert, Casein, Cheddar, Colby, Cream cheese, Edam, Emmenthal, Gouda, Gruyere, Jarisberg, Monterey jack, Munster, Neufchatel, Parmesan, Provolone, Sherbet, Swiss, Whey, Whole Milk.

MEATS & POULTRY	SEAFOOD	EGGS & DAIR
Avoid	Avoid	Avoid
Bacon, chicken, Cornish Hens, Duck, goose, Ham, Heart, Partridge, Pork, Quail.	Anchovy, Barracuda, Beluga, Bluegill bass, Clam conch, Crab, Crayfish, Eel, Frog, Lobster, Lox (smoked salmon), Mussets, Octopus, Oysters, Sea Bass, Shrimp, Snail, Striped bass, Turtle, Yellowtail.	American Cheese, Blue cheese, Ice Cream, String cheese.

BLOOD GROUP "B" FOODS 2

OILS & FATS	NUTS & SEEDS	BEANS & LEGUMES	CEREALS
Highly Beneficial	Highly Beneficial	Highly Beneficial	Highly Beneficial
Olive Oil	NONE	Beans: kidney, lima, navy,	Millet, Oat bran, Oatmeal Price: puffed, rice bran, Spelt.

OILS & FATS	NUTS & SEEDS	BEANS & LEGUMES	CEREALS
Neutral	**Neutral**	**Neutral**	**Neutral**
Cod liver oil, Linseed (flaxseed) oil.	Almond butter, Nuts: almonds, brazil, chestnuts, hickory, litchi, macadamia, pecans, walnuts.	Beans: broad, cannellini, copper, green, java, jacamar, northern, red, snap, string, tamarind, white Peas: green, pods.	Cream or rice, Familla, Farina, Granola, Grape nuts.
Avoid	**Avoid**	**Avoid**	**Avoid**
Canola oil, Corn oil, Cottonseed oil, Peanut oil, Safflower oil, Sesame oil, Sunflower oil.	Nuts: cashews, filberts, pignola (pine), pistachio, Peanuts, Peanut butter, Poppy seeds, Pumpkin seeds, Sesame butter (tahini) Sesame seeds, Sunflower butter, Sunflower seeds.	Beans: aduke, azuki, black, garbanzo, pinto Lentils: domestic, green, red Peas: black-eyed	Amaranth, Barley, Buckwheat, Cornflakes, cornmeal, Cream of wheat, Kamut, Kasha, Rye, Seven-grain, Shredded wheat, Wheat bran, Wheat germ.

BLOOD GROUP "B" FOODS 3

BREADS & MUFFINS	GRAINS & PASTA	VEGETABLES
Highly Beneficial Brown rice bread, Essene bread, Ezekiel bread, Fin Crisp, Milet, Rice cakes, Wasa bread.	**Highly Beneficial** Oat flour, Rice flour.	**Highly Beneficial** beets, Beet leaves, Broccoli, Cabbage: Chinese, red, white. Carrots, Cauliflower, Collard greens, Eggplant, Kale, Lima beans, Mushroom: shitake, Mustard greens, Parsley, Parsnips, Peppers: green, red, jalapeno, yellow, Potatoes: sweet, Sprouts: Brussels, Yams: all types.

BREADS & MUFFINS	GRAINS & PASTA	VEGETABLES
Neutral	Neutral	Neutral
Gluten-free bread, High protein no-wheat bread, Ideal Flat bread, Oat bran muffins, Pumpernickel, spelt bread, Soy flour bread.	Flour: Graham, spelt, white, Pasta: semoline, spinach, Quinoa, Rice: basmati, brown, white.	Arugula, Asparagus, Bamboo shoots, Bok choy, Celery, Chervil, Chicory, Cucumber, Daikon radish, Dandelion, Dill, Endive, Escarole, Fennel, Fiddlehead ferns, Barlic, ginger, Horseradish, Kohlrabi, Leek, Lettuce: Bibb, Boston, iceberg, romaine, mesclun, Mushrooms: abalone, domestic, enoki, Portobello, tree oyster, Okra, Onions: green, red, Spanish, yellow, Potatoes: red, white, Radicchio, Rappini, Rutabaga, Scallion, Seaweed, Shallots, Snow peas, spinach, sprouts: alfalfa, Squash: all types, Swiss chard, Turnips, Water chestnut, Watercress, Zucchini.

BREADS & MUFFINS	GRAINS & PASTA	VEGETABLES
Avoid	Avoid	Avoid
Bagels: wheat, Corn muffins, Durum wheat, Mult-grain bread, 100% rye crisp, Rye Crisp, Rye Vita, What bran muffins, Whole wheat bread.	Buckwheat, kasha, Couscous, flour: barley, rye, bulgur wheat, durum whet, whole wheat, gluten Pasta: artichoke, Noodles: soba, Rice: wild.	Artichoke: domestic, Jerusalem, Avocado, Corn: white, yellow, Olives: black, Greek, green Spanish, Pumpkin, radishes, sprouts: mung, radish, Tempeh, Tofu, Tomato.

BLOOD GROUP "B" FOODS 4

FRUIT	JUICES & FLUIDS	CONDIMENTS
Highly Beneficial	Highly Beneficial	Highly Beneficial
Bananas, Cranberries, Grapes: black, red, concord, green, Papaya, Pineapple, Plums: dark, red, green.	Cabbage, Cranberry, Grape, Papaya, Pineapple.	NONE

FRUIT	JUICES & FLUIDS	CONDIMENTS
Neutral	**Neutral**	**Neutral**
Apples, Apricots, Blackberries, Blueberries, Boysenberries, Cherries, Currants, black, red, Dates, Elderberries, Figs: dried, fresh, Gooseberries, Grapefruit, Guava, Kiwi, Kumquat, Lemons, Limes, Loganberries, Mangoes, Melon: cantaloupe, canang, Crenshaw, Christmas, casaba, honeydew, musk, Spanish, watermelon, Nectarines, Oranges, Peaches, Pears, Plantains, Prunes, Raisins, Raspberries, Strawberries, Tangerines.	Apple, Apple cider, Apricot, Carrot, Celery, Cherry: black, Cucumber, Grapefruit, Orange, Prune, Water (with lemon), Vegetable juice (corresponding with highlighted vegetables).	Apple butter, Jam (from acceptable fruits), Jelly (from acceptable fruits), Mayonnaise, Mustard Pickles: sour, dill, kosher, sweet, Relish, Salad dressing (low-fat, from acceptable fruits ingredients), Worcestershire, sauce.
Avoid	**Avoid**	**Avoid**
Coconuts, Persimmons, Pomegranates, Prickly pear, Rhubarb, Star fruit (carambola).	Tomato.	Ketchup.

BLOOD GROUP "B" FOODS 5

SPICES	HERBAL TEAS	MISC. BEVERAGES
Highly Beneficial Cayenne pepper, Curry, Ginger, Horseradish, Parsley.	**Highly Beneficial** Ginger, Ginseng, Licorice, Parsley, Peppermint, Raspberry leaf, Rose hips, Sage.	**Highly Beneficial** Tea: green

SPICES	HERBAL TEAS	MISC. BEVERAGES
Neutral	Neutral	Neutral
Agar, Anise, Arrowroot, Basil, Bay leaf, Bergamot, Brown rice syrup, Capers, Caraway, Cardamom, Carob, Chervil, Chives, Chocolate, Clove, Coriander, Cream of tartar, Cumin, Dill, Dulse, Garlic, Honey, Kelp, Maple syrup, Marjoram, Mint, Miso Molasses, Mustard (dry), Nutmeg, Oregano, Paprika, Pepper: peppercorn, red flakes, Peppermint, Pimiento, Rice syrup, Rosemary, Saffron, Sage, Salt, Savory, Soy sauce, Spearmint, Sugar: brown, white, Tamarind, Tarragon, Thyme, Turmeric, Vanilla, Vinegar: apple cider, balsamic, white, red wine, Wintergreen.	Alfalfa, Burdock, Catnip, Cayenne, Chamomile, Chickweed, Dandelion, Dong quai, Echinacea, Elder, Goldenseal, Green tea, Hawthorn, Horehound Licorice root, Mulberry, Saint-John's wort, Sarsaparilla, Slippery elm, Spearmint, Strawberry leaf, Thyme, Valerian, Vervain, White birch, White oak bark, Yarrow, Yellow dock.	Beer, Coffee: regular, decaf, Tea black: regular, decaf Wine: red, white.

SPICES	HERBAL TEAS	MISC. BEVERAGES
Avoid	Avoid	Avoid
Allspice, Almond extract, Barley malt, Cinnamon, Cornstarch, Corn syrup, Gelatin: plain, Pepper: black ground, white, Tapioca.	Aloe, Coltsfoot, Corn silk, Fenugreek, Gentian, Hops Linden, Mullein, Red clover, Rhubarb, Senna, Shepherd's purse.	Liquor: distilled Seltzer water, Soda: club, cola, diet other.

BLOOD GROUP "O" FOODS 1

MEATS & POULTRY	SEAFOOD	EGGS & DAIRY
Highly Beneficial	Highly Beneficial	Highly Beneficial
Beef: regular, ground, Buffalo, Heart, Lamb, Liver, Mutton, Veal, Venison.	Bluefish, Cod, Hake, Halibut, Herring, Mackerel, Pike, Rainbow trout, Red snapper, Salmon, Sardine, Shad, Snapper, sole, Striped bass, Sturgeon, Swordfish, Tilefish, White perch, Whitefish, Yellow perch, Yellowtail.	NONE

MEATS & POULTRY	SEAFOOD	EGGS & DAIRY
Neutral	**Neutral**	**Neutral**
Chicken, Cornish hens, Duck, Partridge, Pheasant, Quail, Rabbit, Turkey.	Abalone, Albacore (Tuna), Anchovy, Beluga, bluegill bass, Carp, Clam, Crab, Crayfish, Eel, Flounder, Frog, Gray sole, Grouper, Haddock, Lobster, Mahi-mahi, Monkfish, Mussels, Ocean perch, Oysters, Pickerel, Porgy, sailfish, Scallop, Sea bass, Sea trout, Shark, Shrimp, Silver perch, Smelt, Snail, Squid (Calamari), Turtle, Weakfish.	Butter, Farmer, Feta, goat cheese, Mozzarella, Soy cheese, Soy milk.

MEATS & POULTRY	SEAFOOD	EGGS & DAIRY
Avoid	**Avoid**	**Avoid**
Bacon, Goose, Ham, Pork.	Barracuda, Catfish, Caviar, Conch, Herring (pickled), lox (smoked salmon), Octopus.	American Blue cheese, Brie, Buttermilk, Camembert, Casein, Cheddar, Colby, Cottage, Cream cheese, Edam, Emmenthal, Goat milk, gouda, Gruyere, Ice cream, Jarisburg, Kefir, Monterey jack, Munster, Neufchatel, Parmesan, Provolone, Ricotta, Skim or 2% milk, String cheese, Swiss Whey, Whole milk, Yogurt: all varieties.

BLOOD GROUP "O" FOODS 2

OILS & FATS	NUTS & SEEDS	BEANS & LEGUMES	CEREALS
Highly Beneficial Linseed (flaxseed) oil, Olive Oil.	**Highly Beneficial** Pumpkin seeds, Walnuts.	**Highly Beneficial** Aduke, Aduki, Pinto, Black-eyed.	**Highly Beneficial** NONE
Neutral Canola oil, Cod liver oil, Sesame oil.	**Neutral** Almonds, Almond butter, Chestnuts, Filberts, Hickory, Macadamia, Pecans, Pignola (pine), Sesame butter (tahini), Sesame seeds, Sunflower butter, Sunflower seeds.	**Neutral** Beans: black, broad, cannellini, fava, garbanzo, green, jicama, lima northern, red, red soy, snap, string, white Peas: green, pods.	**Neutral** Amaranth, Barley, Buckwheat, Cream of rice, Kamut, Kasha, Millet: puffed, Rice bran, Rice: puffed, Spelt.

OILS & FATS	NUTS & SEEDS	BEANS & LEGUMES	CEREALS
Avoid	**Avoid**	**Avoid**	**Avoid**
Corn oil, Cottonseed oil, Peanut oil, Safflower oil.	Brazil, Cashew, Litchi, Peanuts, Peanut butter, Pistachios, Poppy seeds.	Beans: copper, Kidney, navy, tamarind Lentils: domestic, green, red.	Cornflakes, Cornmeal, Cream of wheat, Familia, Farina, Grape nuts, Oat bran, Oatmeal, Seven-grain, Shredded Wheat, What bran, wheat germ.

BLOOD GROUP "O" FOODS 3

BREADS & MUFFINS	GRAINS & PASTAS	VEGETABLES
Highly Beneficial Essene Bread, Ezekiel Bread, Sourdough Bread.	**Highly Beneficial** NONE	**Highly Beneficial** Artichoke: domestic, Beet leaves, Broccoli:Chicory, Collard greens, Dandelion, Escarole, Garlic, Horseradish, Kale, Kohirabi, Leek, Lettuce: romaine, Okra, Onions: red, Spanish, yellow, Parsley, Parsnips, Peppers: red, Potatoes: sweet, Pumpkin, Seaweed, Spinach, Swiss chard, Turnips.

BREADS & MUFFINS	GRAINS & PASTAS	VEGETABLES
Neutral	**Neutral**	**Neutral**
Brown Rice Bread, Fin Crisp, Gluten-free bread, Ideal Flat Bread, Millet, Rice cakes, 100 % rye bread, Rye Crisps, Rye Vita, Soy flour bread, Spelt bread, wasa bread.	Barley flour, Buckwheat, Kasha, Pasta: artichoke, Quinoa rice: basmati, wild, brown, white Rice flour, Rye flour, Spelt flou.r	Arugula, Asparagus, Bamboo Shoots, Beets, Bok Choy, Caraway, Carrots, Celery, chervil, coriander, cucumber, Daikon, dill, Endive, Fennel, Fiddlehead Ferns, ginner, Lettuce: Bibb, Boston, iceberg, mescium, Lima Beans, Mushroom: abalone, enoki, Portobello, tree oyster, Olives: green, Onions: green, Peppers: green, yellow, jalapeno, Radicchio, Radishes, Rappini, Rutabaga, Scallion, Shallots, Snow Peas, Sprouts: mung, radish, Squash: all types, Tempeh, Tofu, Tomato, Water chestnut, Watercress, Yams: all types, Zucchini.

BREADS & MUFFINS	GRAINS & PASTAS	VEGETABLES
Avoid	Avoid	Avoid
Bagels: wheat, Corn muffins, Durum wheat, English muffins, High-protein bread, Matzos: wheat, Multi-grain bread, Oat bran muffins, Pumpernickel, Sprouted wheat bread, Whole wheat bread.	Bulgur Wheat Flour, Couscous Flour, Durum Wheat Flour, Gluten Flour, Graham Flour, Oat Flour, Soba Noodles, Past: Semolina, Spinach, Sprouted Wheat Flour, White Flour, Whole Wheat Flour.	Avocado, Cabbage: Chinese, Red, white, Cauliflower, Corn: white, yellow, Eggplant, Mushroom: domestic, shitake, Mustard greens, Olives: black, Greek, Spanish, Potatoes: red, white, Sprouts: alfalfa, Brussels.

BLOOD GROUP "O" FOODS 4

FRUIT	JUICES & FLUIDS	CONDIMENTS
Highly Beneficial	Highly Beneficial	Highly Beneficial
Figs: dried, fresh, Plums: dark, green, red, Prunes.	Black cherry, Pineapple, Prune.	NONE

FRUIT	JUICES & FLUIDS	CONDIMENTS
Neutral	Neutral	Neutral
Apples, Apricots, Bananas, Blueberries, Boysenberries, Cherries, Cranberries, Currants: black, red, Dates: red, Elderberries, Elderberries, Grapefruit, Grapes: red, concord, black, green, Guava, Guava, Kiwi, Kumquat, Lemons, Limes, Loganberries, Mangoes, Melons: canang, casaba, crenshaw, Christmas, musk, Spanish, watermelon, Nectarines, Papayas, Peaches, Pears, Persimmons, Pineapples, Pomegranates, Prickly pear, Raisins, Raspberries, Star fruit (carambola).	Apricot, Carrot, celery, Cranberry, Cucumber, Grape, Grapefruit, Papaya, Tomato water (with lemon), Vegetable juice (corresponding with highlighted vegetables).	Apple butter, Jam (from acceptable fruits), Jelly (from acceptable fruits), Mayonnaise, Mustard, Salad dressing (low-fat from acceptable ingredients), Worcestershire, sauce.

FRUIT	JUICES & FLUIDS	CONDIMENTS
Avoid	**Avoid**	**Avoid**
Blackberries, Coconuts, Melons: cantaloupe, honeydew, Oranges, Plantains, Rhubarb, Strawberries, Tangerines.	Apple, apple cider, Cabbage, Orange.	Ketchup, Pickles: dill, kosher, sweet, sour, Relish.

BLOOD GROUP "O" FOODS 5

SPICES	HERBAL TEAS	MISC BEVERAGES
Highly Beneficial	**Highly Beneficial**	**Highly Beneficial**
Carob, Curry, Dulse, Kelp (bladder wrack), Parsley, Pepper: cayenne, Turmeric.	Cayenne, Chickweed, Dandelion, Fenugreek, Ginger, Hops, Linden, Mulberry, Parsley, Peppermint, Rose Hips, Sarsaparilla, Slippery elm.	Club soda, Seltzer Water.

SPICES	HERBAL TEAS	MISC BEVERAGES
Neutral	**Neutral**	**Neutral**
Agar, Allspice, Almond extract, Anise, Arrowroot, Barley malt, Basil, Bay Leaf, Bergamot, Brown rice syrup, Cardamom, Chervil, Chives, Chocolate, clove, coriander, Cream of tartar, Cumin, dill, Garlic, Gelatin: plain, Honey, Horseradish, Maple syrup, Marjoram, Mint, Miso, Molasses, Mustard (dry), Paprika, Pepper: peppercorn, red flakes, Peppermint, Pimiento, Rice syrup, Rosemary, Saffron, Sage, Salt, Savory, Soy sauce, Spearmint, Sacanat, Sugar: brown, white, Tamari, Tamarind, Tapioca, Tarragon, Thyme, Wintergreen.	Catnip, Chamomile, Don quai, elder ginseng, Green tea, Hawthorn, Horehound, Licorice root, Mullein, Raspberry leaf, Sage, Skullcap, Spearmint, Thyme, Valerian, Vervain, White birch, white oak, bark Yarrow.	Beer, Green tea Wine: red, white.

SPICES	HERBAL TEAS	MISC BEVERAGES
Avoid	**Avoid**	**Avoid**
Capers, Cinnamon, cornstarch, Corn syrup, Nutmeg, Pepper: black ground, white, Vanilla, Vinegar: apple cider, balsamic, red wine, white.	Alfalfa, Aloe, Burdock, Coltsfoot, Corn silk, Echinacea, Gentian Goldenseal, Red clover.	Coffee: regular and decaf Liquor: distilled, Soda: cola, diet other Tea, black: regular & decaf.

COMMON FOODS OR DIETS TO EAT FOR YOUR BLOOD TYPE A, B, AB & O (DOWNLOAD GIDEXFDA APP …)

FOOD ITEMS	ABO BlOOD TYPE			
Foods/Diet List	A	B	AB	O
Skim milk	-	+	/	-
Milk 1-2%	-	+	/	-
Whole milk	-	+	-	-
Soy milk	+	-	/	/
Yogurt (flavored)	/	+	+	-
Yogurt (plain)	/	+	+	-
Cottage cheese	-	+	+	-
Cream cheese	-	/	/	-
Other cheese (cheddar and Swiss)	-	/	/	-
Butter	-	/	-	/
Mayonnaise	-	/	/	-
Ice cream	-	-	-	-
Grapes	/	+	+	/
Prunes	+	/	/	+
Bananas	-	+	-	+
Cantaloupe	/	/	/	-
Honeydew	-	/	/	-
Watermelon	/	+	+	/
Avocado	/	-	-	-
Apple	/	/	/	/
Apple juice	/	/	/	/
Apple butter	/	+	/	/
Orange	-	/	-	-
Orange juice	-	/	-	-
Grapefruit	+	/	+	/
Grapefruit juice	+	/	/	/

Other fruit juices (pineapple etc.)	+	+	/	+
Strawberries	/	/	/	/
Blueberries	+	/	/	+
Peaches	/	/	/	/
Tropical fruits (pineapple etc.)	+	+	+	/
Dried apricot	+	/	/	/
Raisins	/	/	/	/
Dried dates	/	/	/	/
Cranberries	+	+	+	/

FOOD ITEMS (CONTINUED)	A	B	AB	O
Tomatoes	-	-	/	/
Tomato juice	-	-	/	/
Tomato sauce	-	-	/	/
Tofu	+	-	+	/
Green beans	+	/	/	/
Broccoli	+	+	+	+
Cabbage	-	+	/	/
Cauliflower	/	+	+	-
Brussels Sprouts	/	+	/	/
Raw carrots	+	+	/	/
Cooked carrots	+	+	/	/
Corn	/	-	-	-
Peas (green/pod/snow)	+	/	/	/
Mixed vegetables	/	/	/	/
Beans	/	/	/	/
Squash	/	/	/	/
Zucchini	/	/	/	/
Eggplant	-	+	+	/
Yams	-	+	+	/
Garlic	+	/	+	/
Cooked spinach	+	/	/	+
Raw spinach	+	/	/	+
Kale	+	+	+	+
Iceberg	/	/	/	/
Romaine lettuce	+	/	/	+
Celery	+	/	+	/
Green peppers	-	+	-	/
Red peppers	-	+	-	+
Onions (Garnish)	+	/	/	+
Onions	+	/	/	+

Okra	+	/	/	+
Cucumber	/	/	+	-
Pickles	-	/	-	-
Olives	-	-	-	-
Olive oil	+	+	/	+
Beets	/	+	+	/
Asparagus	/	/	/	/

FOOD ITEMS (CONTINUED)	A	B	AB	O
Peanut butter	+	-	+	-
Peanuts	+	-	+	-
Almond butter	/	/	/	/
Almonds	/	/	/	/
Walnuts	+	+	+	+
Egg whites	/	/	/	/
Eggs	/	/	/	/
Bacon	-	-	-	-
Chicken sandwich	/	/	/	/
Chicken/Turkey with skin	/	/	/	/
Chicken/Turkey without skin	/	/	/	/
Pork/Beef hotdogs	-	-	-	-
Chicken/Turkey hotdogs	-	-	-	-
Processed meat sandwich	-	-	-	-
Processed meat	-	-	-	-
Lean hamburger	-	-	-	-
Hamburger	-	-	-	-
Beef/Pork/Lamb Sandwich	/	/	/	/
Pork	-	-	-	-
Beef	-	/	-	+
Liver	-	/	/	+
Chicken liver	/	-	-	/
Other organ meats (heart etc.)	-	-	-	+
Canned tuna	/	/	+	/
Fish cakes	/	/	/	/

Shrimp	-	-	-	/
Dark meat fish (salmon etc.)	+	+	/	/
Other fish (cod etc.)	+	+	+	+
Cold cereal	/	/	/	/
Cooked oatmeal	+	+	+	/
Oat bran	+	+	+	/
Cooked breakfast cereal	/	/	/	/
White bread (flour)	/	/	/	-
English muffins	-	/	/	-
Heavy bread (rye etc.)	/	-	+	/
Dark bread (whole wheat etc.)	-	-	/	-

FOOD ITEMS (CONTINUED)	A	B	AB	O
Bagel	/	/	/	-
Biscuits	/	/	/	/
Brown rice	/	/	+	/
White rice	/	/	+	/
Mac & cheese	-	-	-	-
Pasta (durum wheat flour)	/	-	/	-
Whole grain pasta	/	/	/	/
Wheat germ	-	-	/	-
Barley	/	/	/	/
Bulgur	/	-	/	-
Other grains (couscous etc.)	/	-	/	-
Pancakes	/	/	/	/
French fries	-	-	-	-
Potatoes (scalloped)	-	/	/	-
Potatoes (baked)	-	/	/	-
Potato chips	-	-	-	-
Corn chips	-	-	-	-
Crackers	/	/	/	/
Pizza	-	-	-	-
Tortillas (wheat flours)	-	-	/	-
Coffee	+	/	-	-
Green Tea	+	+	+	+
Black Tea	-	/	-	-
White Wine	/	/	/	/
Red Wine	+	/	/	/
Beer	-	/	/	-
Soda	-	-	-	-
Liquor	-	-	-	-

KEY:

The "+" signs indicate the foods that are Recommended **(Promote wellness)** for the blood groups "A, B, AB, or O" according to the "Blood-Type" Foods. The "-" signs indicate the foods to Avoid for the blood group **(Cause inflammation)**. The "/" signs indicate the foods that are Neutral **(Unknown).**

Dr. Gideon Nnabuike, DNP, APRN, PMHNP-BC, MS, BSN

APPENDIX

BLOOD TYPES AND HEALTH TIPS YOU NEED TO KNOW

1. *** Each human being has one of the four blood types: A, B, AB, and O. Blood types are further classified based on the presence or absence of specific antigens (or proteins) on the surface of red blood cells as A+, A-, and B+, B-, AB+, AB-, O+, and O-, respectively. To Avoid Health Problems, 'You Must Eat Foods According to Your Blood Type.'

2. ***How do we eliminate inflammation and reduce c-reactive protein (CRP)? Depending on your blood type, eat anti-inflammatory foods. Certain foods are linked to higher levels of inflammation. Limiting or avoiding inflammatory foods like refined carbohydrates, fried foods, red meat, and processed meat can help reduce CRP. Eating a heart-healthy diet (e.g., eating plenty of anti-inflammatory foods such as salmon, tuna, and plant-based proteins, avoiding processed meat, consuming omega-3 fatty acids or monounsaturated fatty acids, and including more fresh fruits and vegetables) and reducing stress will prevent CRP.

3. ***" If you want to eat ultra-processed food (UPF), eat it with many fibers or vegetables. Carniola oils are the worst on the market. The blue zones, like Okinawa, Japan, live longer no matter their blood types – they eat sweet potatoes, green tea, fish, and leafy vegetables".

4. *** Eat less seedless watermelon, seedless grape, genetically

modified foods, and vegetables for all blood types. Other examples of foods to avoid are eggs and meat with hormones and chemical preservatives. Hens lay 100 eggs in 24 hours (injected with growth hormone causing dementia and Alzheimer's in humans)), a cow produces 100 gallons of milk in one day (injected with steroids and antibiotics), cow lean meat (Injected with hormones or chicken meat grown in the lab) and meat on store shelves packed with sodium benzoate preservatives with a six-month expiration date.

5. *** Wash your fruits and vegetables to remove pesticides and film coatings. Use apple cider or white vinegar to clean your fruits or veggies; both effectively eliminate bacteria. Then, pour 1 cup of vinegar (237 ml) into the bowl for every 2 cups (473 ml) of cool water.

6. *** "Many foods don't spoil anymore due to biotechnology; plants don't have seeds and become manmade and packed with preservatives."

7. ***" Pineapples, pears, and bananas no longer have extended harvest seasons. "When you control food, you control humans and food specific to blood type.

8. ***" Rice, milk, and meat diets like in Africa die of parasites and autoimmune disease. Hormones and antibiotics are injected into animals, and meats have preservatives and pesticides".

9. ***To have a bowel movement, you must have fiber and water. Drink more water each day!

10. ***" Coconut water has natural minerals to hydrate and stabilize sugar, so it's a no-sweetened beverage." This is good for all blood types.

11. *** The World Health Organization (WHO) states that our food and lifestyle cause 95% of diseases; for example, processed meat is categorized as a level 1 carcinogen. We are taught that there is a pill for every illness. "

12. ***" Alcohol is a toxin; that is why people get intoxicated; if consumed, it goes straight to the liver and brain, lower vitamin B levels."

13. ***" Cigarettes are a level one carcinogen. Medical marijuana causes arrhythmia and does less to cure depression/anxiety and pain".

14. ***" Plant-based foods contain magnesium, which relaxes the body and promotes sleep, and calcium, which relaxes the body, makes you calmer, and pumps blood properly."

15. ***" Start your breakfast with easy-digestible foods like fruits. It requires considerable energy to digest the food eaten, especially protein (meat). "

16. ***" Neuroticism is defined as a personality trait associated with emotional instability, irritability, anxiety, self-doubt, depression, and other negative feelings. Those diagnosed with Neuroticisms are likely diagnosed with dementia and Alzheimer's".

17. ***"Have you drunk a cup of water today? Please drink a glass of warm water immediately after you wake up from sleep each morning." It promotes good metabolism and prevents constipation.

18. ***We recommend speaking with your healthcare provider if you suspect a vitamin D deficiency. They can take bloodwork and recommend the appropriate supplementation dosage based on your current levels. Some common signs of vitamin

D deficiency include Fatigue, Mood changes, Muscle pain and weakness, Reduced immune function, Increased inflammation, Frequent infections, Rickets (in children), and Bone disorders.

19. ***Depending on age, older adults need about 600 to 800 IU of vitamin D daily. You can get this through sun exposure and vitamin-D-rich foods, including salmon, eggs, sardines, liver, white mushrooms, and fortified foods. Supplementing with more isn't always better. Since vitamin D is a fat-soluble vitamin, too much can put you at serious risk for toxicity, including adverse side effects like kidney stones and high calcium levels in the blood. Our experts say it is best not to consume more than 4,000 IU daily, which is the tolerable upper intake level.

20. ***, "Vitamin D is a secosteroid metabolized in the morning sun and can be obtained by eating salmon, egg yolk, and mushrooms. It is always low for Black and African Americans, Asians, Hispanics, and some Caucasians. Keeping Vitamin D levels between 50 and 70 nanograms will cut your chances of getting cancer to 15 percent. It is suggested taking Vitamin D with magnesium together boosts your sleep."

21. ***Can taking vitamin D boost your immune system? Vitamin D is essential for bone health and muscle and nerve functions. It also helps the immune system fight off bacteria and viruses. Vitamin D Supplementation Improves More Than 1.5-fold Survival of Digestive Tract Cancers, Including Colorectal, in Patients with a Cancer-Fighting Immune System. You may need about 2,500 to 5,000 daily; blood and lab work must be checked every three months.

22. ***"How Can I Prevent Clogged Arteries? "Clogged arteries are most often caused by fatty plaque buildup in the walls of the arteries, which reduces blood flow to heart muscle cells.

You can prevent developing clogged arteries by doing the following: • Quit smoking • Eat a healthy diet • Reduce your LDL (bad) cholesterol • Reduce high blood pressure • Lose weight • Exercise. If left untreated, clogged arteries will continue to build plaque, increasing the risk of having a heart attack, heart failure, a stroke, or painful legs from peripheral artery disease".

23. ***"Risk factors for atherosclerosis are lifestyle-related diseases such as diabetes, dyslipidemia, hypertension, smoking, and aging. Although aging is inevitable, eating a well-balanced diet and exercising moderately is important to avoid lifestyle-related diseases. Even if a person has these diseases, it is possible to prevent the progression of arteriosclerosis by controlling blood glucose, cholesterol, and blood pressure appropriately through diet and exercise therapy and medications".

24. ***"The main risk factors for clogged arteries are smoking, excessive consumption of alcoholic beverages, sedentary lifestyle, a diet rich in ultra-processed foods, high blood cholesterol, high blood pressure, diabetes, overweight, and obesity. Therefore, Blood groups A, B, and AB should avoid eating red meat and control these conditions".

25. ***" Cod liver oil provides significant amounts of vitamins A, D, and E and omega-3 fatty acids EPA and DHA. Cod liver oil is pressed from the fish's liver, whereas fish oil is pressed from the fish's body. Cod liver oil includes vitamins A and D and omega-3 fatty acids, whereas fish oil typically contains omega-3 fatty acids only. Regularly consuming omega-3 fatty acids EPA and DHA supports healthy triglyceride levels while benefiting cardiovascular health. The American Heart Association recommends eating two servings of fish (particularly fatty fish like sardines, salmon, herring, and whitefish) per week. A daily serving of cod liver oil would quickly provide the recommended

intakes of omega-3 fatty acids (EPA and DHA). Substituting omega-3 fats for saturated and trans fats can help lower the risk of cardiovascular disease and potentially slow the progression of cognitive decline when consumed regularly. Supplements or foods high in the retinol form of vitamin A (like some cod liver oils) should be consumed carefully due to the possibility of ingesting too much regularly. Consult a qualified healthcare practitioner before eating large amounts of retinol."

26. ***What foods are good for your prostate? Bok choy (bought in Vietnamese stores), broccoli, Brussels sprouts, cauliflower, cabbage, and kale are high in essential vitamins, minerals, and antioxidants to help reduce inflammation and maintain a healthy prostate. These vegetables also contain phytochemicals that are known to prevent the growth of cancer cells. Natural remedies, such as soy, green tea, pygeum, grass pollen, and saw palmetto, may help naturally shrink the prostate in some people. Other strategies, such as behavioral changes, may also help to shrink the prostate naturally. Citrus: Oranges, lemons, limes, and grapefruits are all high in vitamin C, which may help to protect the prostate gland.

27. ***What foods cleanse the gallbladder? In most cases, a gallbladder cleanse involves eating or drinking a combination of olive oil, herbs, and some fruit juice over several hours. Proponents claim that gallbladder cleansing helps break up gallstones and stimulates the gallbladder to release them in stool. But some specific vegetables and fruits help you dissolve your gallstones. Foods that contain Vitamin B, Vitamin C, and calcium are best for patients with gallbladder stones. Citrus fruits like lemon and orange, different types of Bell peppers, and leafy Vegetables help reduce gallstones.

28. ****" Drinking one tablespoon of apple cider vinegar with water before a meal can help manage blood sugar after eating. The vinegar can help not only type 2 diabetics but also pre-diabetics and healthy individuals wanting to eat healthier. It works to increase sensitivity and lower blood sugar. Some reports show that apple cider vinegar may reduce HbA1c after 8-12 weeks of use. "

29. *** "Sinus infections, or sinusitis, occur when the nasal passages become inflamed and filled with mucus. This condition can cause discomfort and difficulty breathing. Viral or bacterial infections, allergies, or structural abnormalities in the nasal passages commonly cause sinus infections. They can last for a few days to several weeks, depending on the severity and the underlying cause. This inflammation or blockage/inflamed sinus in older adults may cause ringing in the ear or ears (tinnitus)".

30. ***Calcium deficiency, or hypocalcemia, occurs when you don't have enough calcium in your blood. Calcium is essential in organs like bone health, muscle contraction, nerve transmission, calming effect, and blood clotting. Not getting enough calcium could cause symptoms like muscle cramps, changes in mood, and skin or hair dryness. Muscle cramps and spasms in the hands, feet, or face, tingling and numbness in the hands and feet, weakness, Fatigue, twitching or tremors, and difficulty speaking or swallowing.

31. ***Drinking green or match tea in moderation benefits kidney health. However, consuming green tea in extract form (supplement) in high quantities could cause damage to the liver and kidneys; it is recommended to use caution when ingesting green tea in a concentrated extract format. Limit the intake of green tea extract to less than 800 milligrams per day, as doses above this amount may pose health risks. Matcha tea is low in

calories, averaging three calories per serving. It contributes to weight loss and helps by preventing diseases and delaying aging.

32. ***What is the healthiest fish to eat? Alaskan salmon, Cod, Herring, Mahi-mahi, Mackerel (other than king mackerel), Perch, Rainbow trout, Sardines, Striped bass, Tuna (other than bluefin and bigeye tuna), especially canned light tuna, Wild Alaskan pollock, and Arctic char

33. *** According to One Medical, the following Fish are healthy to eat and have minimal environmental impact: troll-caught or pole-caught albacore tuna from the U.S. or British Columbia, wild-caught salmon from Alaska, farmed oysters, wild-caught sardines from the Pacific Ocean, farmed rainbow trout, and tank-farmed freshwater coho salmon from the U.S.

34. ***Does exercise lower blood pressure? In other words, exercise can lower one's blood pressure - especially in individuals who do it often. In all blood types, "those who exercise regularly tend to have lower blood pressure than those who live more sedentary lifestyles. One caveat is that exercise doesn't lower blood pressure while working out because it increases during exercise to supply your muscles with the additional oxygen and blood flow they need. But once your heart rate returns to normal following the workout session, the benefits of prolonged exercise can be experienced.

35. ***Study shows how fasting may reduce inflammation in all blood types. Many swear that trendy fasting diets are keeping them slimmer and healthier. They may now have some science to back that up. British researchers at the University of Cambridge believe they've uncovered the processes that cause fasting to lower bodily inflammation. Long hours without eating appear to trigger a rise in a blood chemical called arachidonic

acid, which has anti-inflammatory properties, reports a team led by Clare Bryant of Cambridge's Department of Medicine. Bryant's team is focused on what scientists call the "inflammasome" -- the cellular "alarm" system by which the body defends itself from injury or illness, triggering inflammation. "What's become apparent over recent years is that one inflammasome in particular -- the NLRP3 inflammasome -- is essential in several major diseases such as obesity and atherosclerosis, but also in diseases like Alzheimer's and Parkinson's disease, many of the diseases of older age people, particularly in the Western world," Bryant explained.

36. *** Significant variations in blood pressure could be an indication that a person is at risk of a heart attack or stroke, a study has suggested. Researchers warned there is an urgent need for "new practical ways" to assess variability and are exploring technology for patients to track their blood pressure over time. They found variations in systolic pressure – the higher of the two numbers used to measure blood pressure and the force at which your heart pumps blood around your body – was a "strong predictor" of heart attacks, strokes, or atrial fibrillation. This condition causes an irregular and fast heartbeat. The cohort was split into three, based on average systolic and blood pressure variability. For example, among the group with an average systolic blood pressure of less than 140 mmHg, those with the most significant variability, 16%, were more likely to have a heart attack or stroke. It can increase the risk of heart attacks and stroke, as well as the likeliness of heart and kidney failure and vascular dementia.

37. *** Vitamin C is needed more by smokers or vapers and people exposed to secondhand smoke. The primary food sources of vitamin C are citrus fruits such as oranges, lemons, tangerines,

acerola, passion fruit, strawberries, pineapple, and kiwi. Studies have found that vitamin C can be an excellent agent to protect agricultural products against fungal contamination and reduce the associated damage due to its adaptation to the human body.

38. *** Keeping one's blood pressure levels in the normal range is essential. Blood pressure that's too high (hypertension) can cause severe headaches and blurred vision and may even lead to heart disease or stroke. Blood pressure that's too low (hypotension) can cause symptoms like light-headedness, confusion, and nausea.

39. *** Why does blood pressure change as we age? **Per the National Institutes of Health (NIH) National Institute on Aging, normal blood pressure for most adults is defined as a systolic pressure of less than 120 and a diastolic pressure of less than 80, written as 120/80. This matters because "as we age, our arteries lose elasticity and essentially become stiffer, which can affect how much blood the heart needs to push in the same way that blowing into a balloon that has not been stretched beforehand takes much more effort than blowing into it after you have stretched it.**

40. ***Peanut oil is rich in vitamin E, an antioxidant that offers many protective benefits against chronic disease. But it also contains pro-inflammatory omega-6 fatty acids that may increase the risk of certain diseases. So, you must consume it in moderation or choose another oil like olive. Peanut oil may also be prone to oxidation, so it is not a good choice as a cooking oil.

41. *** Cottage cheese is a great choice when suffering from diverticulosis. That derives from cottage cheese being high in protein and calcium, having no fiber, being soft, and being

more accessible if you have any side effects from diverticulosis. Dairy products like cottage cheese and Greek yogurt are generally considered safe for people with diverticulosis. These fiber-free foods are high in protein and calcium and can be easy to get down if you're having trouble with solid foods. However, if you're experiencing discomfort, it's best to avoid dairy and speak with your doctor or dietitian to see if they suit you.

42. *** Is cooking foods with aluminum pots safe or wrapping hot food in foil? It's best to use a barrier, like parchment paper, when cooking or storing such foods. Prevent over-cooking: If using aluminum foil to wrap or cover food for baking or grilling, be cautious not to overcook the food. Excessive heat and prolonged contact with aluminum can cause it to leach into the food. What are the signs of aluminum toxicity? Liver stenosis and nephrotic syndrome kidney disease are manifestations of aluminum toxicity, and the brain-respiratory system can also be severely damaged, followed by aluminum poisoning. Memory loss, tremors, jerking, and death are essential manifestations of brain injury.

43. ***How are vitamins absorbed in the gut? The fat-soluble vitamins A, D, E, and K are absorbed from the intestinal lumen using the exact mechanisms to absorb other lipids. In short, they are incorporated into mixed micelles with other lipids and bile acids in the lumen of the small intestine and enter the enterocyte mainly by diffusion. What inhibits vitamin absorption? Lifestyle factors, including stress, medication, diet, caffeine, alcohol, and even exercise, may negatively affect the absorption of nutrients.

44. ***While there may be many reasons for chronic fatigue, needing vitamin D is one of them. Changes in sleep and not sleeping well can also be symptoms to pay attention to. Number eight,

"Is feeling depressed and sad." Slow wound healing can signal that you're down on the sunshine vitamin. He reports that this can be a big problem because spine wounds need to heal, too. Finally, being sick often and having frequent infections might require supplementation.

45. *** Vitamin D toxicity - Even if you have nine out of 10 symptoms, seeing your doctor and checking your levels before beginning a new supplement is always advisable. While lacking this vitamin can lead to unwanted health conditions like rickets or osteoporosis, too much can also have detrimental effects. According to the Mayo Clinic, signs of a problem include "The main consequence of vitamin D toxicity is a buildup of calcium in your blood (hypercalcemia), which can cause nausea and vomiting, weakness, and frequent urination. Vitamin D toxicity might progress to bone pain and kidney problems, such as the formation of calcium stones." Which is why getting tested should be the first step.

46. ***Zinc deficiencies are rare in the average population. Still, people with gastrointestinal diseases, sickle cell disease, vegetarians, alcoholism, or pregnant or breastfeeding are all more likely to lack the proper amount of zinc either due to malabsorption or not getting enough in their diet. Zinc is vital for appropriate immunity and immune system function, so people lacking in zinc may get sick more often or from more minor issues. Symptoms of deficiency may also include upset stomachs and diarrhea.

47. ***Iron deficiencies are widespread in children, people who menstruate, vegetarians, and vegans. Children are likely to have iron deficiencies unless given iron-rich foods regularly. Menstruation causes iron deficiencies because of the regular loss of

blood that occurs. Because meat is an excellent source of iron, many vegetarians and vegans do not have enough iron in their diets. Iron deficiencies can also be caused by anemia when your body does not have enough red blood cells to carry enough oxygen around your body. If you have an iron deficiency, you might experience weakness, fatigue, brain fog, a weakened immune system, or even raised ridges in your nails.

48. ***Iodine is essential for a properly functioning metabolism and thyroid. It involves many hormone processes, so an iodine deficiency can quickly worsen premenstrual symptoms. Iodine is necessary for many important body functions. It is also essential for proper brain function and development, and iodine deficiencies in children have been linked to mental and developmental abnormalities.

49. *** A vitamin B12 deficiency is widespread in vegetarians and vegans. Mainly, vegans will not receive adequate B12 from their diet because it is found only in meat, seafood, eggs, and milk products. Vegetarians are still very likely to have a deficiency as well. If vegetarians, and especially vegans, do not supplement with a plant-based supplement form of vitamin B12, they will be deficient. B12 is necessary for proper blood formation and brain and nerve functioning, so everyone must get an adequate amount of B12, whether through eating animal products or taking a plant-based supplement.

50. *** Calcium is necessary for proper bone health. We've all been told to drink milk for strong bones, and that is because milk is a good source of calcium. You can also find it in dark green vegetables and sardines or anchovies. Calcium is transported around your body through your blood, storing the extra calcium in your bones. So, when the stores are running low, your

bones are weaker, which makes them more prone to breaking, especially with age. Brittle nails are also a sign of calcium deficiency.

51. ***There are two forms of dietary vitamin A, one of which can be found in meat and animal products, while the other is in fruits and vegetables like carrots and leafy greens. Both are necessary for our bodies, but in Western culture, we typically have more than enough in our diets, so deficiencies are not as expected. However, these deficiencies are widespread in developing countries, and Indian women are more likely to suffer from a vitamin A deficiency. Vitamin A is necessary for skin, teeth, bones, and cell membranes. Deficiency in vitamin A can lead to blindness.

52. ***While it is uncommon for people nowadays to get a vitamin C deficiency as severe as scurvy, not getting enough daily vitamin C can still interfere with some of your normal bodily functions. Vitamin C is essential for hormone and amino acid formation, involving nearly every body part. Vitamin C and iron work hand in hand; if you do not consume enough vitamin C, you are more likely to have an iron deficiency. No matter how much iron you drink, it will not be absorbed without vitamin C. So, if you show signs of an iron deficiency, it could be a vitamin C deficiency.

53. *** Magnesium is another essential mineral for your bones and teeth. A deficiency may be the result of drug use, poor digestion, or just not getting enough magnesium in your diet. Hospital patients are prone to magnesium deficiencies. Symptoms of a deficiency can include migraines, cramps, and irregular heartbeat. Magnesium is not only necessary for your bones and teeth but also crucial for helping your muscles relax. Epsom salt (magnesium sulfate) is suitable for a relaxing bath.

54. *** Omega 3 fatty acids are the "good" fats found in nuts, seafood, and avocados, but really, what makes them the "good" fats is that we get a disproportionate amount of omega six fatty acids in the standard Western diet. Omega 6 fatty acids and omega three fatty acids need to be paired in a balanced ratio, so an omega three deficiency could mean you are also getting way too much omega six compared to omega 3 in your diet. Either way, signs of an omega-three deficiency are problems with hair, skin, and nails, fatigue, brain fog, joint pain, menstrual irregularities, and cardiovascular concerns. These concerns can also be present if you are deficient in iron or calcium, which can pair with an omega-3 deficiency.

55. ***Folate deficiencies have the most severe consequences in the early stages of pregnancy since this can cause neural tube defects in the growing fetus. This can cause a baby to be born with a congenital disability called spina bifida, which is why all sexually Active people who could become pregnant are strongly encouraged to be on a multivitamin with folate. However, folate isn't just important for those who can become pregnant and who do not eat enough nutritious meals.

56. **** Which is the healthiest bread to eat? Whole Grain Bread. Brown bread is made using whole wheat, with its outer covering intact. This makes brown bread more nutritious and fiber richer than white bread. Brown bread contains vitamins B-6 and E, magnesium, folic acid, zinc, copper, and manganese. Thanks to their high fiber and nutrient content, breads made with whole grains, including whole wheat, are generally the go-to healthiest breads dietitians recommend. "Most people need more fiber, so finding high-fiber bread is often a good idea. If you are in blood group 0, have BP, or are diabetic, eat sourdough bread.

Is sourdough the healthiest bread? According to a 2019 editorial review in Aging Clinical and Experimental Research, sourdough bread has a lower glycemic index and glycemic load than white bread and whole-wheat bread that is not fermented. Whole-wheat sourdough is higher in fiber, lowering the strain it puts on your blood glucose.

57. *** Beets are rich in dietary nitrates, a compound with blood-pressure-lowering effects. Beets are rich in nitrates, and "raw beet juice lowers blood pressure and contains vitamins, minerals, and antioxidants. Its efficacy is backed up by a randomized clinical trial showing improvements in blood pressure and indices of systemic inflammation."

58. ***Berries are full of antioxidants, great for anti-aging benefits. "Enjoy fresh blueberries in your oatmeal (not for Blood group O). Some people enjoy berry juice, and it has been found that consuming berries lowers both systolic blood pressure. You can add berries to oatmeal or as a snack to get more antioxidants into your diet.

59. *** Green tea and coffee have blood pressure-lowering effects, "although tea's benefits are more apparent: While both black and green tea lowered blood pressure in a recent meta-analysis of all available trial data, green tea was more productive. "Coffee is more controversial as acute effects of coffee and caffeine intake include increased blood pressure. However, moderate coffee consumption over an extended period shows cardiovascular risk and blood pressure-lowering effects.

60. ***Tomato juice has been shown to improve systolic and diastolic blood pressure. "I enjoy tomatoes on my salad. I have also found that many people enjoy tomato juice. The unsalted

variety, "as has been shown to reduce both blood pressure and cholesterol in clinical studies, contains vitamin C and antioxidants.

61. ***Formulas to reduce pain, liver detox, and wellness: ginger roots, seven small turmeric roots. 2 tablespoons of ground cayenne and black pepper, and seven pieces of cut pineapple. Drink 3 to 4 tablespoons as needed each day. Consult your doctor if you're pregnant or nursing babies or have other underlying health issues.

62. *** To boost your immunity, drink the following before breakfast two times a week: Take four cloves of garlic root, one purple onion, and two whole lemons, cut them into pieces, and boil for 10 minutes. Consult your doctor if you're pregnant, nursing, or have other underlying health issues.

63. ****These two ingredients will heal your pain if you are experiencing knee, back, or any pain. Get on avocado seed (ripped or uniped avocado). Cut into sizes or pieces and put in a visible or transparent plastic bottle container. Add 70% alcohol, cover the cut avocado seeds in the bottle, and shake slowly. Then, use a cotton swab to soak the mixture and apply it evenly to the areas in which you are experiencing pain. Repeat as desired until the pain goes away.

64. *** Drinking a glass of Chayote squash juice mixed with one lime juice first in the morning will cure several diseases, including high blood pressure (HBP). Also, boiling water for 10 minutes, add 1-2 gm Hibiscus Flower (Zobo), sift and drink, do not add anything else to the flavonoid -red color. Drinking a glass each day with a meal (taking Zobo drink in an empty may cause sudden drop of your HBP to fatal level and may lead to digestive discomfort), Zobo reduces menstrual pain, manage high

blood pressure, improve appetite, lower cholesterol, anti-cancer properties, antidepressant agent, prevents anemia, uses as a laxative, protects the heart against cardiovascular diseases and promote weight loss. Do not take Zobo drink after 24 hours.

65. *** Incorporate more raw garlic (2 cloves a day, once a week), pumpkin seeds (unsalted palm of hand whole snacks), beets, and carrots into your diet. These natural remedies, long used to combat parasites, can empower you to take control of your health. In a study, a mixture of honey and papaya seeds successfully cleared the stools of parasites in 23 out of 30 subjects. Additionally, staying hydrated by drinking plenty of water can aid in flushing out your system.

66. *** Pumpkin seeds, a powerhouse of nutrients, can significantly benefit your health. They are rich in healthy fats, antioxidants, and fiber, promoting gut flora and heart and liver health. Their benefits are amplified when combined with other healthy grains like flax seeds. Moreover, the seeds of pumpkins and many other vine crops are believed to contain a deworming compound called cucurbitacin, which has been used to expel tapeworms and roundworms in domestic livestock species for years.

67. *** Incorporate Oregano in your meals or drink tea; it contains chemicals that might help reduce cough, arthritis, allergies, sore throat, and asthma. It might also help with digestion and fight against some bacteria and viruses. People use oregano for wound healing, parasite infections, and many other conditions, but no good scientific evidence supports these uses.

68. ***A serving of walnuts is roughly 1 ounce or about ¼ cups. However, some studies suggest eating 1.5 to 2 ounces may be beneficial to reap the health benefits walnuts offer. With this in mind, and considering walnuts are calorically dense, Kalloo rec-

ommends aiming for a handful daily to reap their many health benefits. Walnuts are packed with minerals, vitamins, and fiber. They're also a good source of protein and healthy fats, linked to keeping you full longer and helping you control your cholesterol levels.

69. ** It is not too late to reset your biological life clock (every human lifespan ticks daily). To achieve optimal health and longevity, do these four things: 1. Exercise daily, 2. Eat plant-based foods, 3. Avoid eating UPF (ultra-process foods), and 4. Reduce inflammation.

REFERENCES

1. Du Toit, G., Roberts, G., Sayre, P. H., et al. (2015). Randomized trial of peanut consumption in infants at risk for peanut allergy. *New England Journal of Medicine*, 372(9), 803-813.

2. Greer, F. R., Sicherer, S. H., Burks, A. W., & Committee on Nutrition and Section on Allergy and Immunology. (2008). Effects of early nutritional interventions on the development of atopic disease in infants and children: The role of maternal dietary restriction, breastfeeding, timing of introduction of complementary foods, and hydrolyzed formulas. *Pediatrics*, 121(1), 183-191.

3. LeBovidge, J. S., Strauch, H., Kalish, L. A., & Schneider, L. C. (2008). Assessment of psychological distress among children and adolescents with food allergy. *Journal of Allergy and Clinical Immunology*, 122(6), 268-273.

4. Sicherer, S. H., & Sampson, H. A. (2014). Food allergy: Epidemiology, pathogenesis, diagnosis, and treatment. *Journal of Allergy and Clinical Immunology*, 133(2), 291-307.

5. Turner, P. J., Baumert, J. L., Beyer, K., et al. (2015). Can we identify patients at risk of life-threatening allergic reactions to food? *Allergy*, 70(2), 149-160.

Printed in the USA
CPSIA information can be obtained
at www.ICGtesting.com
LVHW010858210924
791651LV00012B/476